ENDPAPER
Charles Lindbergh's *Spirit of St. Louis* approaches a spectacularly illuminated Paris in the final minutes of his stirring transatlantic flight in 1927. French artist Paul Lengellé created this *gouache* rendering expressly for *The Pathfinders*.

THE PATHFINDERS

Other Publications:

THE GOOD COOK
THE SEAFARERS
THE ENCYCLOPEDIA OF COLLECTIBLES
THE GREAT CITIES
WORLD WAR II
HOME REPAIR AND IMPROVEMENT
THE WORLD'S WILD PLACES
THE TIME-LIFE LIBRARY OF BOATING
HUMAN BEHAVIOR
THE ART OF SEWING
THE OLD WEST
THE EMERGENCE OF MAN
THE AMERICAN WILDERNESS
THE TIME-LIFE ENCYCLOPEDIA OF GARDENING
LIFE LIBRARY OF PHOTOGRAPHY
THIS FABULOUS CENTURY
FOODS OF THE WORLD
TIME-LIFE LIBRARY OF AMERICA
TIME-LIFE LIBRARY OF ART
GREAT AGES OF MAN
LIFE SCIENCE LIBRARY
THE LIFE HISTORY OF THE UNITED STATES
TIME READING PROGRAM
LIFE NATURE LIBRARY
LIFE WORLD LIBRARY

FAMILY LIBRARY:
HOW THINGS WORK IN YOUR HOME
THE TIME-LIFE BOOK OF THE FAMILY CAR
THE TIME-LIFE FAMILY LEGAL GUIDE
THE TIME-LIFE BOOK OF FAMILY FINANCE

THE EPIC OF FLIGHT

THE PATHFINDERS

by David Nevin

AND THE EDITORS OF TIME-LIFE BOOKS

TIME-LIFE BOOKS, ALEXANDRIA, VIRGINIA

Correspondents: Elisabeth Kraemer (Bonn); Margot Hapgood, Dorothy Bacon, Lesley Coleman (London); Susan Jonas, Lucy T. Voulgaris (New York); Maria Vincenza Aloisi, Josephine du Brusle (Paris); Ann Natanson (Rome). Valuable assistance was provided by Nakanori Tashiro, Asia Editor, Tokyo. The editors also wish to thank: Janny Hovinga (Amsterdam); Martha Mader (Bonn); Enid Farmer (Boston); Brigid Grauman (Brussels); Nina Lindley (Buenos Aires); Sandy Jacobi (Copenhagen); Lucia Apcar (Detroit); Peter Hawthorne (Johannesburg); Pat Stimpson (London); Diane Asselin (Los Angeles); John Dunn (Melbourne); Felix Rosenthal (Moscow); Carolyn T. Chubet, Miriam Hsia, Christina Lieberman (New York); Mary Martin, John Scott (Ottawa); M. T. Hirschkoff (Paris); Mimi Murphy (Rome); Robert Plaskin (St. John's); Janet Zich (San Francisco); Carol Barnard (Seattle); Peter Allen (Sydney); Eiko Fukada (Tokyo); Nancy Friedman (Washington, D.C.).

THE AUTHOR

David Nevin covered aviation as a reporter in Texas and has published nine books on subjects ranging from Micronesia to American political leaders. *The Pathfinders* is his fifth Time-Life book. The four previous books were part of The Old West series: *The Soldiers, The Expressmen, The Texans* and *The Mexican War.*

THE CONSULTANT for The Pathfinders

Richard P. Hallion Jr. is Curator of Science and Technology at the National Air and Space Museum, Washington. The author of four books on aviation topics as well as many articles on the history and technology of aviation, Dr. Hallion also serves as a professor of history at the University of Maryland.

THE CONSULTANTS for The Epic of Flight

Melvin B. Zisfein, the principal consultant, is Deputy Director of the National Air and Space Museum, Washington. He received degrees in aeronautical engineering from the Massachusetts Institute of Technology and has contributed to many scientific, technological and historical publications. He is an Associate Fellow of the American Institute of Aeronautics and Astronautics.

Charles Harvard Gibbs-Smith, Research Fellow at the Science Museum, London, and the Keeper-Emeritus of the Victoria and Albert Museum, London, has written or edited some 20 books on aeronautical history. In 1978 he served as the first Lindbergh Professor of Aerospace History at the National Air and Space Museum, Washington.

Dr. Hidemasa Kimura, honorary professor at Nippon University, Tokyo, is the author of numerous books on the history of aviation and is a widely known authority on aeronautical engineering and aircraft design. One plane that he designed established a world distance record in 1938.

For information about any Time-Life book, please write:
Reader Information
Time-Life Books
541 North Fairbanks Court
Chicago, Illinois 60611

First printing.
Published simultaneously in Canada.
School and library distribution by Silver Burdett Company, Morristown, New Jersey.

Library of Congress Cataloguing in Publication Data
Nevin, David.
The pathfinders.
(The Epic of flight; v. 2)
Bibliography: p.
Includes index.
1. Aeronautics—United States—History.
2. Aeronautics—Flights. I. Time-Life Books.
II. Title. III. Series: Epic of flight; v. 2.
TL521.N48 629.13'092'4 [B] 79-19867
ISBN 0-8094-3256-0
ISBN 0-8094-3255-2 lib. bdg.

CONTENTS

1	**The lure of the mighty Atlantic**	**15**
2	**Revving up for Paris**	**51**
3	**Tragedy and triumph**	**77**
4	**The long reach to Australia**	**109**
5	**Globe-girdler from Oklahoma**	**139**
	Appendix: Great flights of the pathfinders	170
	Acknowledgments	172
	Bibliography	172
	Picture credits	173
	Index	174

Daring vanguard of a golden age

"Aeronautics," wrote the Russian aviation pioneer Igor Sikorsky, "was neither an industry nor a science. It was a miracle." And so it must have seemed in its early days, not only to the many who watched fearful and flabbergasted from below but to the relative few who, like the young Sikorsky, actually soared aloft.

The miracle was achieved a hundred times over by the audacious pilots of aviation's infancy. It was these venturers who were the first to ask: Can I get there from here in an airplane? They often answered the question in machines that men a decade later would have considered instruments of suicide. Their dreams, and their achievements, seem modest only in hindsight. Before great oceans could be vaulted, after all, the English Channel had to be crossed. In 1909 it *was* crossed, at its narrowest breadth, by a mustachioed Frenchman named Louis Blériot *(right)* in a skeletal craft that he designed himself. Miracle indeed. The 540-pound monoplane carried Blériot, his crutches—made necessary by an earlier accident—and not a single navigational instrument.

After the Channel, the Alps were conquered, then the Pyrenees, and in 1913, the Mediterranean Sea. Great cities were linked, at least tenuously, by men flying against the clock, and each other, for glory and rich prizes: London and Paris, then Paris and Rome, Paris and Cairo. Miracles all—and they were by no means always safely accomplished. Blériot and his contemporaries had to contend not only with undependable flying machines but also with every caprice of weather and wind; on occasion *(page 10)* they even had to fight off the birds whose domain they had invaded.

The advent of war in 1914 ended the brief era of the first pathfinders, but their deeds had contributed an inspiring prelude to the golden age of aviation.

The Tricolor and the British Red Ensign welcome France's Louis Blériot as his 25-hp monoplane surmounts the cliffs of Dover to land near Dover Castle (background) on July 25, 1909. The 23½-mile crossing from Calais took 36½ minutes. The inset portrait of Blériot, like those of several of the fliers on the following pages, has been added to the original illustration, which appeared on the cover of Le Petit Journal a month after the historic flight.

A Paris crowd in August 1910 cheers Alfred Leblanc's first-place finish in the Circuit de l'Est, a cross-country race to six French cities that lasted 10 days. Leblanc covered the 491-mile course in a total flying time of 12 hours 1 minute 1 second.

Peruvian Georges Chavez is pulled mortally injured from the wreckage of his Blériot airplane after crashing on his trailblazing flight over the Alps in September of 1910. His last words, "Higher, ever higher," became the motto of his country's Air Force.

Eugène Gilbert, crossing the Pyrenees in the 1911 Paris-to-Madrid race, fires a shot from his revolver to ward off an angry eagle. Gilbert survived the bird's attack but was forced out of the race by engine trouble. Two years later he won the Paris-to-Rome race.

André Beaumont, the first pilot to fly from Paris to Rome, arrives over the Italian capital on May 31, 1911. A remarkably steady flier, Beaumont went on to win two major long-distance races that same year: the Circuit of Europe and the Circuit of Britain.

Jules Vedrines, the winner of the 1911 Paris-to-Madrid race, heads for a crash on the railroad tracks near Épinay-sur-Seine, France, in the contest's 1912 rerun. Vedrines survived, and in December of 1913 he became the first aviator to fly from Paris to Cairo.

Roland Garros of France became the first pilot to fly across the Mediterranean, just making it from St. Raphaël in southern France to Bizerte, Tunisia, in September 1913. His plane had enough fuel for eight hours of flight; the journey lasted 7 hours 53 minutes.

1

The lure of the mighty Atlantic

With the end of World War I, peacetime aviation seemed poised for an exhilarating leap forward. The great challenge—and the great obstacle—that dominated the thinking of the men who built airplanes and those who flew them was the Atlantic Ocean. Like some menacing beast from an ancient myth, it lay in wait between the Old World and the New, an intolerable barrier, both real and psychological.

To begin with, the Atlantic was immense—almost four times wider at its narrowest point than the longest distance that had yet been flown over water. It produced weather of nightmarish proportions: gales that could reduce air speed to a walk and blow a plane miles off course in an hour, or billows of freezing fog that rose to the heavens and could coat a plane with ice in minutes, forcing it into the sea. And the best route across this treacherous pond lay far north of the shipping lanes, making rescue unlikely in the event of a crash.

Such perils were enough to frighten away the prudent. But to a few men the dangers were irresistibly seductive. Within the fraternity of aviation there was unspoken agreement that one day soon the Atlantic barrier would be overcome. The fliers and builders were, after all, immediate heirs of the prewar aviators who had proved *(pages 6-13)* that, given flying machines that were little more substantial than motorized kites, daring men could leap between cities, soar over lofty mountain ranges and cross expansive bodies of water that had seemed impassable until the first airmen showed the way.

Certainly before 1919 the Atlantic was beyond the aviator's reach. But the Great War had changed all that. During the War the contending armies had pressed, forcefully and successfully, for more powerful engines and better-designed planes to fly heavier payloads for greater distances. After the War ended, these new aircraft—and succeeding generations of ever-more-capable ones—pointed the way to a golden era of epic flights that eventually would leave no continent untouched, no sea or ocean unspanned.

The men who accepted the challenge knew that they flew at the fringe of aeronautical knowledge, each pressing to do what no one had done before. As the era began, airplanes, though capable of marvelous feats, had not yet become so reliable that marvelous feats were commonplace. Navigation aids were rudimentary at best. Once the takeoff roll began, these adventurers in the cockpit were on their own, with little to

John Alcock and Arthur Whitten Brown, who were leaders in the quest to connect the continents of the world by air, stand immortalized in stone at London's Heathrow International Airport.

rely on but their finely tuned skills as pilots and the engines that droned along reassuringly up ahead. Like Columbus and Magellan, who centuries earlier had first crossed the great oceans, these men were pathfinders—shrinking by air the world that seafarers had enlarged under sail. And their first goal was the Atlantic.

At daybreak on April 28, 1910, John Alcock, an 18-year-old apprentice engineer, watched with thousands of other flying enthusiasts outside Manchester, England, as an adventurous Frenchman, Louis Paulhan, bounced safely to earth in his Farman biplane. On landing, Paulhan claimed one of aviation's first rich rewards: the £10,000—approximately $50,000—offered by Lord Northcliffe, owner of *The Daily Mail,* to the first pilot to fly the 186 miles northward from London to Manchester within 24 hours. Alcock, an eager lad with a loud laugh, a ruddy face and a crop of hair the color of ginger, had loved speed and airplanes since his boyhood, and Paulhan's feat stirred him to act. Soon afterward, Alcock took a job as mechanic at one of England's first schools for training men in the skills of flying.

Alfred, Lord Northcliffe, depicted at age 30 in an 1895 issue of Vanity Fair, was an early believer in the usefulness of aviation and promoted its development by having his newspaper, The Daily Mail, offer a prize in 1913 for the first transatlantic flight.

The aviation school was located at Brooklands—a race track turned airdrome about 20 miles southwest of London—and it was run by Maurice Ducrocq, a flier from France, which was the European center of aviation in 1910. At Brooklands toiled airplane builders whose names and products would soon become famous in World War I—Bristol, Avro, Sopwith, Martinsyde, and ultimately the most important for Alcock, the recently established aviation division of Vickers Ltd., the famous armaments manufacturer.

As Ducrocq's sole mechanic, Alcock was on the road to the life he had dreamed of. He repaired airplanes and off duty he haunted the Blue Bird Restaurant, endlessly talking aviation with other young men in greasy whipcords and leather jackets. Alcock himself learned to fly in two hours by sitting behind Ducrocq in flight and resting his novice hands on Ducrocq's experienced ones.

Alcock soloed in an old Farman machine pushed along by a propeller in the rear; he soon entered a weekend competition at nearby Hendon airdrome and won a race. *Aeroplane,* a magazine that chronicled the feats of the early aviators, ran his picture under the heady caption, "Mr. Jack Alcock—the latest crack pilot."

For all the enthusiasts of the Blue Bird café there were air derbies to win, distance and altitude records to break. But these weekend events paled in comparison to the new challenge—and reward—offered in 1913 by Lord Northcliffe's *Daily Mail:* another £10,000 prize, this time for the first flight across the Atlantic.

At the Blue Bird, airplane builders, fliers and mechanics chewed eagerly on this provocative idea. Clearly, no plane then flying could perform the feat. But Lord Northcliffe's intent was to stimulate aeronautical advances. And what strides such a flight would demand! The shortest distance across the Atlantic Ocean—some 1,880 miles—lay between Newfoundland, the eastern prominence of North America, and Ireland. A flight of such great distance either would demand an impossible number of refueling and maintenance stops in the water, or for a nonstop flight, would require enormous gasoline tanks and engines that could operate faultlessly for 20 to 30 hours, roughly 10 times the average achieved in 1913.

Before any pilot or builder could rise to Lord Northcliffe's challenge, World War I intervened. Alcock joined the Royal Naval Air Service and as a flight lieutenant flew bombing missions against the Turks, who shot him down in the waning months of the War and took him prisoner. In captivity, he talked endlessly to his cellmates of his plans to fly across the Atlantic. Demobilized as a captain in the Royal Air Force, which had incorporated the Naval Air Service in 1918, Alcock returned to Brooklands looking for a job. At Vickers Ltd., he managed to talk himself into the perfect assignment: to fly their latest aircraft in quest of Northcliffe's Atlantic prize.

The works manager at the Vickers plant, a Scotsman improbably named Maxwell Muller, led Alcock into the assembly shed to have a look at the plane. There, half finished, stood a Vickers Vimy, a speedy bomber that was named for a town in northern France near the site of a Canadian victory. The Vimy reflected all the impressive advances in aeronautics that had been made during the War. It was a big, handsome biplane with a two-man cockpit and a wingspan of slightly more than 67 feet. Each of its two reliable 12-cylinder Rolls-Royce Eagle Mark VIII engines was capable of generating 360 horsepower. Once the bomb racks were removed, a forward gunner's cockpit omitted and new tanks installed to nearly double the original fuel capacity, the Vimy would be ready to go. Although 865 gallons of gasoline in the fuselage would make the plane 1,000 pounds overweight (gasoline weighs more than six pounds per gallon), Vickers Ltd. was confident that the Vimy could not only get off the ground but, at a cruising speed of approximately 100 miles per hour, stay in the air long enough to cross the ocean.

With Alcock enrolled as pilot, Vickers had only to find a navigator. A few weeks later, a quiet man in a neat Royal Air Force uniform limped into Max Muller's office leaning on a walking stick. His name was Arthur Whitten "Teddie" Brown and he was an engineer looking for a job. In the interview that followed, Muller happened to mention aerial navigation. It was like opening a floodgate. Brown's reticence vanished as he talked eagerly—and knowledgeably—of sextants, star sightings and drift calculations.

Brown had been born of American parents and raised in Manchester. In 1914 he had relinquished his United States citizenship to join the British Army. He became an aerial observer, then crashed behind German lines, so injuring his left leg that he would never walk normally again. Brown, a prisoner of war like Alcock, had spent his months of incarceration mastering navigation with books procured through the Red Cross. Could he, Muller asked, navigate a plane across the Atlan-

Royal Aero Club of the United Kingdom,
3, CLIFFORD STREET, LONDON, W. 1.

Telegraphic Address: "Aerodom, London."
Telephone: Regent 1327-8-9.

"DAILY MAIL" £10,000 PRIZE.
Cross-Atlantic Flight.

(Under the Competition Rules of the Royal Aero Club.)

The Proprietors of the "Daily Mail" have offered the sum of £10,000 to be awarded to the aviator who shall first cross the Atlantic in an aeroplane in flight from any point in the United States, Canada, or Newfoundland to any point in Great Britain or Ireland, in 72 consecutive hours. (The flight may be made either way across the Atlantic.)

Qualification of Competitors.—The competition is open to persons of any nationality not of enemy origin, holding an Aviator's Certificate issued by the International Aeronautical Federation and duly entered on the Competitors' Register of the Royal Aero Club.

No aeroplane of enemy origin or manufacture may be used.

Entries.—The Entry Form, which must be accompanied by the Entrance Fee of £100, must be sent to the Secretary of the Royal Aero Club, 3, Clifford Street, London, W.1, at least 14 days before the entrant makes his first attempt.

No part of the Entrance Fee is to be received by the *Daily Mail*. All amounts received will be applied towards payment of the expenses of the Royal Aero Club in conducting the competition. Any balance not so expended will be refunded to the competitor.

Starting Place.—Competitors must advise the Royal Aero Club of the starting place selected, and should indicate as nearly as possible the proposed landing place.

All starts must be made under the supervision of an Official or Officials appointed by the Royal Aero Club.

Identification of Aircraft.—Only one aircraft may be used for each attempt. It may be repaired en route. It will be so marked before starting that it can be identified on reaching the other side.

Stoppages.—Any intermediate stoppages may only be made on the water.

Towing.—Towing is not prohibited.

Start and Finish.—The start may be made from land or water, but in the latter case the competitor must cross the coast line in flight. The time will be taken from the moment of leaving the land or crossing the coast line.

The finish may be made on land or water. The time will be taken at the moment of crossing the coast line in flight or touching land.

If the pilot has at any time to leave the aircraft and board a ship, he must resume his flight from approximately the same point at which he went on board.

GENERAL.

1. A competitor, by entering, thereby agrees that he is bound by the Regulations herein contained or to be hereafter issued in connection with this competition.

2. The interpretation of these regulations or of any to be hereafter issued shall rest entirely with the Royal Aero Club.

3. The competitor shall be solely responsible to the officials for the due observance of these regulations, and shall be the person with whom the officials will deal in respect thereof, or of any other question arising out of this competition.

4. A competitor, by entering, waives any right of action against the Royal Aero Club or the Proprietors of the *Daily Mail* for any damages sustained by him in consequence of any act or omission on the part of the officials of the Royal Aero Club or the Proprietors of the *Daily Mail* or their representatives or servants or any fellow competitor.

5. The aircraft shall at all times be at the risk in all respects of the competitor, who shall be deemed by entry to agree to waive all claim for injury either to himself, or his passenger, or his aircraft, or his employees or workmen, and to assume all liability for damage to third parties or their property, and to indemnify the Royal Aero Club and the Proprietors of the *Daily Mail* in respect thereof.

6. The Committee of the Royal Aero Club reserves to itself the right, with the consent of the Proprietors of the *Daily Mail*, to add to, amend or omit any of these rules should it think fit.

1st February, 1919.

For Entry Form, see over.

Lord Northcliffe's contest rules reflect the short range of aircraft before World War I, permitting a pilot to set down on the ocean, to board a ship, even to have his plane towed, provided that he resumed his flight from the point where he landed.

tic? He could indeed. It was the very puzzle he had posed for himself, and solved, in prison camp.

Muller led Brown out to the works and showed him the Vimy. "That's our bus," he announced, "and there's the man who's going to fly her. Come and meet him."

Brown was instantly impressed. Jack Alcock, dressed in overalls and a tweed jacket, radiated an insouciant confidence. Within minutes he and Brown were sketching transatlantic routes in chalk on the shop floor. Afterward, as they were inspecting the plane, Alcock noticed Brown's walking stick.

"You won't be needing that," he said, with a grin that cemented their partnership. "We're flying, not imitating Moses." There was only one problem: Brown was due to get married soon. That night Teddie Brown broke the news to his fiancée. They would have to delay the wedding because he was going to fly the Atlantic.

By April 1919 the prize for a transatlantic flight—Northcliffe's, now augmented by offers from a private businessman and a tobacco company—had grown to £13,000, and a competition had developed in which Alcock and Brown were late starters. There were five other contenders, four British and one Swedish. Alcock was a friend of two of the competing pilots. Frederick Raynham at 26 looked barely out of his teens, but he had learned to fly in 1911 at Brooklands and that same year took charge of the Sopwith airplane company's flying school. The other friend, one of Raynham's first students, was a young Australian named Harry Hawker, a lean, smallish man with dark curly hair and a quick smile. Four days after his first flying lesson, Hawker had soloed. Back injuries he had suffered in numerous plane and car accidents had kept him out of military service in World War I but had not prevented him from test flying 283 aircraft in a two-year period.

Major J. C. P. Wood's Shamrock is towed ashore after splashing into the Irish Sea on April 18, 1919. Upon his rescue—and seeing that the waterlogged Shamrock would never fly again—Wood declared, "The Atlantic flight is pipped."

Such gifted pilots were naturally attracted to the Atlantic competition, and so was Britain's aircraft industry. Like Vickers, other firms were preparing to compete for what they regarded as a potentially huge postwar market in commercial aviation; winning the race across the Atlantic would be a convincing demonstration of any airplane's reliability and safety.

By the time Alcock joined Vickers, Sopwith had under construction an aircraft of new design, which the company pointedly christened the *Atlantic.* A single-engined biplane in which the pilot sat beside and a little behind the navigator, the *Atlantic* had a feature that was unique among the competing aircraft. Following takeoff, the plane would be streamlined by dropping its landing gear—a belly skid would be used for landing. The resulting seven-mile-per-hour gain in air speed would enable a single Eagle Mark VIII engine to pull the plane along at 100 miles per hour for 3,000 miles on 330 gallons of fuel. And in case the plane fell into the ocean, the upper part of the Sopwith's fuselage was designed to serve as a lifeboat. To fly the *Atlantic,* Sopwith turned to its most competent pilot, Harry Hawker.

The Martinsyde Company also produced an entrant—a somewhat smaller single-engined two-seater. Though powered by a Rolls-Royce Falcon mustering only 285 horsepower, it could fly 110 miles per hour because it was lighter, and even with wheels attached, more streamlined than the Sopwith. Carrying 393 gallons of fuel, the Martinsyde aircraft had a range of 2,750 miles. Freddie Raynham would pilot the plane and he counted on its greater speed to make him a winner if the contest developed into a neck-and-neck race.

Both firms engaged navigators who had been trained, like almost all aerial navigators of the era, on the sea. Sopwith chose Lieutenant Commander Kenneth Mackenzie Grieve, known as Mac, a Scotsman who had joined the Royal Navy when he was 14. Now, at the age of 39, he was a tall, quiet, almost cadaverous man who had had no flying experience whatsoever until Hawker gave him his first flight training. Martinsyde's navigator was Captain C. W. Fairfax Morgan, like Alcock a veteran of the Royal Naval Air Service. Fax Morgan claimed to be a direct descendant of Henry Morgan, the pirate. Like Teddie Brown, he had been shot down over France, but his luck was worse than Brown's; his left leg had been amputated and replaced by an artificial one that was made of cork. But then, as Alcock had said, none of them intended to walk across the Atlantic.

One other big name in British aviation, Handley Page, Ltd., fielded an entry. Its new V/1500 bomber—like the Vimy, built to bomb Berlin—was the biggest plane flying. Powered by four Rolls-Royce Eagle engines, it would carry a crew of two plus a commander. In charge would be Admiral Mark Kerr, who at the age of 55 was old enough to be John Alcock's father.

Two contestants were eliminated early in the game. In the United States, a Swedish-American pilot named Hugo Sunstedt entered a biplane of his own design. Called *Sunrise,* it flew well enough to chance the ocean, and Sunstedt was an experienced flier. But during a test flight in February of 1919, another pilot spun the *Sunrise* into the sea off Bayonne, New Jersey.

On April 8, Major J. C. P. Wood became the second competitor to drop out; with little preparation and less thought, he took off from Eastchurch, England, in a Short Brothers seaplane and headed west. Twenty-two miles beyond Holyhead, Wales, his engine quit and he ditched in the Irish Sea, from which he was duly rescued.

Wood and his navigator were fortunate that their engine conked out when it did. In flying from east to west, they were bucking head winds that would have exhausted their fuel supply before they reached Newfoundland, and they would have gone down in the icy North Atlantic. To avoid that fate, the other competitors had decided to transport their aircraft to Newfoundland by sea. Starting from there, they could expect to ride a tail wind all the way to Ireland. By May 1, Hawker and the Sopwith *Atlantic,* Raynham and the Martinsyde *Raymor* (named for the pilot and for navigator Morgan), Admiral Kerr with the Handley Page

V/1500 and their support crews were all either en route to Newfoundland by ship or already there.

Only Alcock and Brown were still in England, completing the Vimy and gathering equipment and supplies—fuel, navigation instruments, even an electrically heated jacket for Brown to wear under his flying suit, which came from Burberrys, the celebrated London haberdasher. On Good Friday they had attached the last wire brace to the Vimy and tried it out. Alcock was delighted with the maiden flight. "It's a piece of cake," he remarked of the challenge ahead. "All we have to do is keep the engines going and we'll be home for tea." Finally they crated the plane—fuselage in one box, wings in two others, engines in a fourth—for the voyage to Newfoundland by freighter. Alcock and Brown took the passenger ship *Mauretania* to Halifax, Nova Scotia, then boarded a train for the two-day trip by rail and ferry to St. John's, Newfoundland. They arrived on May 13, well ahead of the Vimy. It was midnight, in the middle of a violent storm, when they beat on the door of the Cochrane Hotel. Eventually it was opened by Agnes Dooley, the proprietor's sister-in-law, who helped to run the establishment.

"We're the Vickers Party," Alcock announced. "We've come to fly the Atlantic."

"Lord save us," said Miss Dooley, "more flying people. The hotel's full of you already. Is the whole world gone daft?" But she took them in, and the next morning at breakfast Hawker and Raynham and their crews gave the Vimy fliers a raucous welcome.

The immediate problem, as Alcock and Brown soon learned, was to find an airfield. Newfoundland had none in 1919. Hawker observed later that he "knew from maps that Newfoundland was the last place to look for spacious landing ground, but if anything, the maps seemed to flatter the country. The hilly terrain made level fields of any size extremely rare, and to take off, the heavily laden planes would need an unobstructed run of at least 300 yards. The best of the makeshift sites had already been taken by the early arrivals. Raynham and Morgan occupied a meadow near Quidi Vidi lake, a half mile from St. John's. Hawker had claimed an L-shaped field of 40 acres near Mount Pearl, six miles away. Its longer leg, he thought, might be just long enough for the Sopwith. However, the field lay at a windy elevation and was hemmed in by a disturbing row of trees. Neither of the strips would be long enough to accommodate the overweight Vimy.

Admiral Kerr, who arrived at about the same time as Alcock and Brown did, took the Handley Page to an isolated site near Harbour Grace, 60 miles away. A reporter's description of this takeoff strip showed just how difficult the conditions were: "It wasn't one field but a series of gardens and farms with rock walls between them. All of these had to be removed, as did three houses and a farm building. A heavy roller, drawn by three horses and weighed down with several hundred pounds of iron bars, eliminated the hummocks. The result, after a month, was a bumpy aerodrome."

Shy Arthur Whitten Brown (left) and gregarious John Alcock, despite differing personalities, make an ideal team for the Atlantic race. Both grew up in Manchester, England, and both developed a long-smoldering desire to fly the Atlantic.

Brown spent his days in St. John's asking where he might find level ground, while Alcock drove an old Buick touring car around the farm roads in a similar search. The only fields that they were able to locate were planted, and the cost to purchase and destroy the crops would be prohibitive.

It began to look as though Alcock and Brown would be the last contestants to start. They had no airfield and their airplane was still at sea. The assembly of Admiral Kerr's plane was progressing at Harbour Grace, where the admiral was living comfortably as the house guest of Robert Reid, the wealthy builder of Newfoundland's only railroad. At St. John's, in the meantime, Hawker and Raynham had their planes ready. Only the weather held them back, and each was like a greyhound on a leash. Every morning they went to the naval station at Mount Pearl, only to hear the same report from the meteorologists: bad weather over the North Atlantic. Every morning Agnes Dooley at the Cochrane Hotel prepared sandwiches and Thermoses for the aviators to take aloft with them if the weather improved. And every night they returned to the hotel, laughing and joking but increasingly tense. The two pilots watched each other so carefully that each finally agreed to alert the other before going. The bad weather was all the more frustrating because the sun seemed to be shining everywhere else in the world. Mac Grieve, Hawker's navigator, received a message from the women

Awaiting a break in the bad weather in St. John's, Newfoundland, friendly rivals (from left) Kenneth Mackenzie Grieve, Frederick Raynham, Harry Hawker and Fairfax Morgan clown on the steps of their lodgings, the Cochrane Hotel.

in the cable office of the British War Mission in New York: SIR, DO BUCK UP AND START. WE CANNOT STAND THE SUSPENSE MUCH LONGER. BEST OF LUCK FROM TWO CABLETTES.

With little to do but fret about the weather, the two crews amused themselves with practical jokes, sometimes choosing as victims the reporters who had gathered in St. John's from London and New York. Among the newsmen was a gullible fellow from a New York paper. One night he asked Morgan how he proposed to navigate to Europe. With a straight face, the pirate's descendant whispered: "Well, that's a bit of a secret, but I like you, so here it is. I've got six carrier pigeons whose home is Brooklands. After we have flown about 300 miles I'll release one. The bird will head straight for Brooklands and we'll just alter course to his direction." Releasing a bird every 300 miles would get them over Ireland, he said, "and then we just follow the chart." History does not reveal whether the newsman actually sent the story to his paper.

Hawker and Raynham slipped a smelly codfish into the bed of another reporter, who had kept everyone awake by typing at night. They poured water into the hollows of the deep leather chairs in the hotel smoking room and awaited victims. The jokes were no doubt abetted by judicious tippling from the limited local supply of liquor. In Newfoundland prohibition was the law in 1919, the result of a measure that had been passed during the War. But the airmen had a great fan in the local physician, who prescribed "tonic waters" that were dispensed from the back door of the drugstore. Moreover, a dozen cases of comforting spirits, packed in a box marked aviation spares, had accompanied Raynham's plane from England.

May 18, for a change, was windy and bright. Hawker ordered the Sopwith fueled, and Raynham's men began filling the Martinsyde's tanks. Both pilots haunted the weather station, where the picture was uncertain. In fact, foul weather was building, but it would be seven years before ships at sea began making regular weather reports by radio, and the deteriorating situation was not evident at St. John's. Hawker saw conditions as "not yet favorable—but possible." At noon he decided to go and informed Raynham. By midafternoon the *Atlantic's* tanks were brimming and its engine had been tested. Agnes Dooley's sandwiches were stored, a sack of letters was loaded aboard and seals were affixed to the plane to prove, when it landed in Ireland, that it had taken off from Newfoundland. Hawker and Grieve struggled into watertight suits, shook hands with many of the crowd that had gathered, and climbed into the cockpit.

Now came the first test. Like the other contenders, the Sopwith had been converted into an overweight flying gasoline tank. The takeoff would be hazardous. Hawker had made two 900-mile test flights before leaving England, but neither had required taking off with a full load of gasoline. To complicate matters further, the field was soft and bumpy.

"Tell Raynham I'll greet him at Brooklands," Hawker shouted above

A storm-battered crossing by flying boat

On the evening of May 16, 1919, three United States Navy Curtiss flying boats, designated *NC-1, NC-3* and *NC-4,* lifted off from Trepassey Bay, Newfoundland, and headed southeastward over the darkening ocean. Though their goal was to cross the Atlantic, they were not competing for the £10,000 prize offered by Lord Northcliffe. Their flight was to be a businesslike air-sea maneuver: They intended to use a chain of 41 American destroyers posted as navigation aids to help guide them from Canada to Portugal, with a refueling stop in the Azores.

At first, the 14-ton flying boats cruised smoothly over the whitecapped Atlantic. "Our four 400-horsepower Liberty engines hummed monotonously," recalled Lieutenant Walter Hinton, copilot of the *NC-4*. "The destroyers were passed as regularly as railway stations."

Toward morning, however, good weather gave way to heavy rains and fog. The flying boats, then within 350 miles of the Azores, were thrown off course and separated. Running low on fuel, the *NC-1* and the *NC-3* were forced to land on the sea. Their crews eventually reached Azorian ports. The six men of the *NC-1* abandoned their plane and were taken on board a Greek merchant ship; the *NC-3's* crew turned the plane's nose into the wind, and using the tail assembly as a sail, rode with the storm for 62 hours. Finally they turned on the engines and taxied into port.

Meanwhile, the *NC-4* had flown on, hoping for a break in the clouds. "Luck was with us," reported Hinton. "The fog split open and we saw the northern tip of Fayal Island." The *NC-4* landed in Horta harbor 15 hours and 13 minutes after leaving Newfoundland. Ten days later, the *NC-4* lifted off again, and at dusk on May 27, to the sound of cannon salutes from Portuguese warships, the American flying boat alighted at the mouth of the Tagus River near Lisbon. Lieutenant Commander Albert C. Read, navigator of the *NC-4,* radioed his superiors: "We are safely on the other side of the pond. The job is finished."

Its lower wing tip washed away, the NC-3 rides out an Atlantic storm with a crewman hanging from a strut to help balance the plane.

the idling engine. Then he threw a salute and sent the Sopwith lumbering across the field. It was 3:40 p.m. local time. The plane lurched and shuddered across the soft ground and ran a full 300 yards before it lifted off. Up and up it inched, just clearing the ominous line of trees. Hawker could not resist a last turn above his rival at Quidi Vidi lake. The Sopwith headed for the coast, jettisoned its landing gear and in six minutes was out of sight.

Raynham was unperturbed. His faster Martinsyde could still overtake the Sopwith. Two thousand spectators gathered at the airfield to watch him take off. The crowd was in a merry mood despite the presence of danger. A group of girls twitted navigator Morgan about his urine relief hose, which they noticed trailing from the leg of his Burberrys flying suit.

The wind at Quidi Vidi blew across their path instead of into it. The crosswind would reduce lift, but if Raynham was to catch up with Hawker, he could not wait for the wind to change direction. Two hours after Hawker had departed, the boyish-looking Raynham waved from his cockpit and advanced the throttle. A silence fell over the crowd as the plane started to move, its engine roaring. The heavy plane ran 100, 200, 300 yards on the ground. Then a bump tossed it into the air. The Martinsyde sagged along for another hundred feet, drifting sideways in the crosswind; then it fell back to earth. Its landing gear sheared off and the nose dug in with a crash. Horror-stricken, the crowd ran across the field to help. Raynham crawled out but Morgan had to be lifted from the cockpit.

Alcock and Brown were returning from another fruitless search for a field when a motorist hailed them.

"Hawker left this afternoon," the man shouted.

"And Raynham?"

"Machine smashed before he could get it off the ground."

They found Raynham at the hotel, his head bandaged. Morgan lay in

Curious onlookers surround Freddie Raynham's Martinsyde Raymor after he and navigator Fairfax Morgan crashed on takeoff in their attempt to cross the Atlantic. Morgan was injured, so Raynham tried again with a new navigator—and cracked up again. Raynham then quit the contest and returned to Britain by ship.

The tail of the Atlantic, forced down by a faulty radiator, bobs in the mid-ocean swell. Pilot Harry Hawker and navigator Mac Grieve escaped injury; the wrecked plane and what was to have been the first transatlantic airmail—including a soggy letter to His Majesty King George V (bottom)—were recovered four days later.

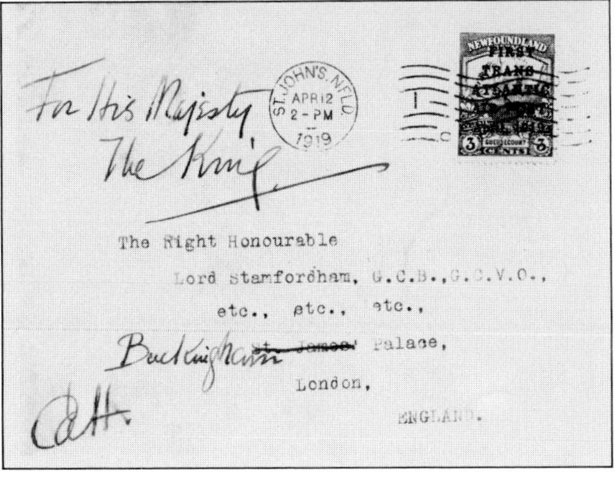

For His Majesty
The King

ST. JOHN'S, NFLD.
APR 12
2-PM
1919

NEWFOUNDLAND

The Right Honourable
Lord Stamfordham, G.C.B., G.C.V.O.,
etc., etc., etc.,
Buckingham ~~St. James~~ Palace,
London,
ENGLAND.

the hospital. That evening, awaiting word of Hawker's expected triumph, Raynham offered the Vimy men his field at Quidi Vidi. At least they could assemble their machine there when it arrived; afterward they would have to move: They would need a longer field to take off fully loaded for the transatlantic flight.

Hawker's Sopwith was equipped with a radio, and the naval station at Mount Pearl was standing by for one of the 24 commercial vessels sailing near Hawker's route to relay his reports of the flight. None came. All night the airmen sat by the Cochrane Hotel's telephone. The jokes died down. At dawn the men began to worry. Hawker's fuel, they estimated, would last for about 22 hours. By afternoon it had become obvious that the Sopwith must be down at sea. Flags at St. John's were lowered to half-mast.

The week that followed was long and sad. The sea had obviously swallowed Hawker and Grieve. At the hospital the one-legged Morgan learned that he would lose an eye. His flying days were over. Then, on Sunday, May 25, Fred Memory, a reporter for *The Daily Mail,* bounded into the hotel, a cable in hand. "They're safe," he shouted. "Harry and Mac landed in Scotland this morning!"

But they had not arrived by air. Earlier that morning a Danish tramp steamer, the *Mary,* had arrived off the coast of an island in the Hebrides and sounded her siren. With flag signals she reported: "Saved hands Sop aeroplane." Fluttering flags from a Navy communications station responded: "Is it Hawker?" "Yes." Then the station transmitted the news to the world.

In the midst of the ensuing celebration at the Cochrane Hotel, Alcock sent an urgent message to Vickers Ltd. at Brooklands: Find out what happened. Soon the details flooded in. First, the Sopwith's radio had failed. The weather had deteriorated steadily, and after four hours the plane was laboring through heavy rain squalls, with clouds towering higher than 15,000 feet. Then Hawker noticed the engine temperature climbing dangerously. Steam began to seep from the radiators; it froze into drops of ice on Hawker's goggles.

Shouting to each other above the racket in the open cockpit, Hawker and Grieve concluded that rust or bits of solder in the radiator must have clogged the water filter. Unless they could somehow jar the particles loose, the radiator water sooner or later would boil away and the engine would seize. Hawker cut the engine, dived 3,000 feet, then pulled the nose up sharply, hoping the maneuver would clear the water filter. The temperature dropped a few degrees but soon was climbing again.

The situation was ironic. "The whole damned Atlantic was beneath us," Hawker said later. "We could have let down a bucket and a rope for water if we had had a bucket and a rope." Again and again he shut off the engine and let the Sopwith dive. Once when he tried to restart the engine, nothing happened. Grieve frantically pumped gasoline from an auxiliary tank as the plane dropped. At 100 feet the engine caught. Hawker "gave her a good mouthful of throttle and she roared away."

All the climbing caused the Sopwith to burn up fuel much faster than Hawker had anticipated. Nine hours from St. John's, he and Grieve had consumed half their gasoline, but they had traveled less than half the distance to Ireland. At dawn, with the radiator steaming steadily, Hawker knew that they would never make it. They had to come down. Grieve estimated that by now they were near the Atlantic shipping lanes, so Hawker flew close to the stormy surface, weaving back and forth in search of a vessel.

After two suspense-filled hours nursing an engine that they expected to die at any moment, they saw a ship through a rain squall. They flew alongside at bridge level and fired a red Very light as a distress signal. Then Hawker set the plane down in a trough between waves. Hastily the two fliers unshipped the lifeboat section of the fuselage and lowered it into a sea that was whipped to 12-foot crests by a brewing gale. Hawker and Grieve bobbed helplessly in their tiny craft for 90 drenching minutes before the crewmen of the ship could get a boat to them. Had Hawker ditched a half hour later, the ship's skipper said, the worsening weather would have put him and Grieve beyond help. By then however, they were safe and dry aboard the *Mary,* drinking aquavit and coffee. But they could not report that they were safe; the ship had no radio. (Amazingly, the *Atlantic* did not sink. Ten days after the crash, a passing ship salvaged the wreckage and returned it to England, where it was displayed in Selfridges, a London department store.)

Off the coast of Scotland, Hawker and Grieve transferred to a British destroyer that took them to London. The two airmen were received as heroes. *The Daily Mail* announced a consolation prize of £5,000, and they were feted everywhere. In St. John's, Alcock observed glumly to Brown that England's "hands are so blistered clapping Harry Hawker that we'll be lucky to get a languid hand"—assuming that they ever got off the ground in Newfoundland.

The huge crates containing the Vimy arrived on May 26, the day after the news of Hawker's rescue and 16 days after Admiral Kerr and the Handley Page crew had arrived at Harbour Grace. A local drayman named Lester, who had hauled the Sopwith and the Martinsyde to their takeoff points, took charge. "Don't you worry, Skipper," he boomed at Alcock. "We'll get your flying machine up to Kiddy Viddy even if we have to knock a few houses down and build a bridge." A fence or two did have to come down, some houses were slightly chipped, and when the crates jammed between bridge parapets the structures had to be disassembled and later rebuilt—but Lester delivered the plane as he had promised.

The Vimy was too big to fit into the hangar tent at the lake so, like Admiral Kerr's Handley Page 60 miles away, it had to be assembled in the open. Canvas screens stretched between posts blocked the worst of Newfoundland's cold spring wind, but frequent rain forced the crew of mechanics and riggers who had accompanied the Vimy from England

Putting the Vimy together again

The Vickers Vimy in which John Alcock and Arthur Brown would attempt to cross the Atlantic finally arrived in Newfoundland—packed into 13 wooden crates. A crew of technicians from England waited to reassemble the plane on the bleak shore of Quidi Vidi lake *(below)*.

Working against time, the crew first erected A-frame hoists to support parts of the Vimy during the assembly process, which is traced on the following pages. The fuselage was raised so the center sections of the lower wing, with the undercarriage attached, could be bolted to it. Then engine nacelles were secured to struts fastened above the wheels, and the remaining wing sections and tail were added. Last, while some mechanics routed control cables to the wings and tail, others installed the Rolls-Royce engines and connected the throttles and fuel lines.

Miserable weather plagued the mechanics as they labored. The Vimy was too large to fit into any available shelter, so the men had to work outdoors, protected from biting gales only by canvas windbreaks. Soldering irons, heated in a fire, sometimes cooled before they could be used and rain often brought work to a complete halt.

Yet in 14 days the plane was ready for a test flight. Even at the Vickers factory in England, that time would have been a record; at Quidi Vidi, it was little short of a miracle.

A tent at Quidi Vidi sheltered mechanics preparing to reassemble the Vimy but was too small to hold the plane itself.

The 50-foot-long crate containing the Vimy's fuselage arrives from the dock at St. John's on a horse-drawn wagon.

Mechanics manhandle the 120-pound center section of the Vimy's upper wing to secure it to upright struts (below).

The crew backs the partially finished Vimy into position for completing the wing assembly (above, right).

A mechanic crouched on a suspended plank bolts a strut to the right wing tip before it is attached to the plane.

The assembly finished, a satisfied crew filters fuel for the Vimy through a funnel lined with copper gauze to trap impurities.

to stop work and cover the plane with tarpaulins. It was brutal, exhausting labor, but the Vimy soon took shape.

For all his explorations of the countryside, Alcock still had not found a field suitable for takeoff with the Vimy fully fueled, but once more Lester, the drayman, saved the day. He had a pasture in mind that might do, and he and the pilot drove out to see it. The field measured only 300 yards, but beyond it lay 200 yards of reasonably open land. Boulders would have to be moved, trees felled, a stone dike taken down and a ditch filled, but the row of meadows satisfied Alcock. "We'll call it Lester's Field," he said. "The first transatlantic aerodrome!"

For three days all available hands worked to smooth the ground. Brown, though his game leg hurt so that he could not pull on overalls without help, toiled alongside the rest. Even soft-palmed journalists bent to the task. As they labored, they sang a Newfoundland work song with an improvised chorus:

> Oh, lay hold Jackie Alcock, lay hold Teddie Brown,
> Lay hold of the cordage and dig in the groun'
> Lay hold of the bow-line and pull all you can,
> The Vimy will fly 'fore the Handley Page can.

The new field was ready on Sunday, June 8, 1919. Looking it over, Alcock said to his navigator, "It'll do, Teddie, it'll do. But I hope we only have to use it once!"

They would have to use it quickly. On that same day, the first warm Sunday of a tardy Newfoundland spring, they heard Admiral Kerr's Handley Page drone serenely overhead. For an awful moment they feared that the old sailor and his crew were setting out for Ireland. But no; the plane turned back toward Harbour Grace. The flight was only a trial run. Unknown to Alcock and Brown, the Handley Page had developed a problem with one of its radiators, and repairs would delay its takeoff for Ireland for several days.

The Vimy flew the next day, lifting off from Quidi Vidi with just enough gasoline to reach Lester's Field after a test flight. On landing at the new airstrip, Alcock gunned the right engine to swing the plane to a stop just short of a fence. MACHINE ABSOLUTELY TOP-HOLE, he cabled to Vickers. And so it seemed, except for the radio, which gave Brown an electric shock. Even more jolting were the sight of the Handley Page wheeling overhead on a second test flight and the discovery that the Vimy's fuel supply was useless, contaminated with a gummy residue that Alcock feared would clog his Rolls-Royce engines. Raynham, whose Martinsyde was not yet repaired, generously offered his gasoline to Alcock, who gratefully accepted. But Raynham's offer proved unnecessary; Max Muller arrived from Vickers with an unexpected supply of fresh fuel.

Alcock was restless; it was time to go. But a gale blew in on June 10 and 11, postponing the flight. An exasperating cable came from Britain: WEATHER PERFECT HERE. PLEASE CABLE REASON FOR NON-START. Disgruntled over the bad weather in Newfoundland and the seemingly interminable series of delays, normally easygoing Alcock lost his temper for the first time, snapping at his ground crew. Brown, to calm himself, went fishing with a fellow angler he had met in Newfoundland.

On June 13 the weather over the Atlantic seemed to be improving, though winds were strong at St. John's. Despite the wind, Alcock had the Vimy fueled with gasoline hand-pumped from drums and filtered through fine copper mesh. One of the riggers tacked a horseshoe under Alcock's seat for luck.

It required most of the morning to fill the tanks. As the plane took on fuel, its weight increased and the wheels sank deeper into the turf. A mechanic clutched Brown's arm and pointed urgently at the undercarriage. A shock absorber had broken and the plane was sagging. It took all afternoon to empty the gasoline tanks so that the plane could be lightened enough to make repairs. The pilots returned to the hotel to rest, while crewmen worked all night by the light of automobile head lamps and paraffin flares to fix the Vimy and pump the fuel back into the tanks for an early takeoff.

On the morning of the 14th, everything was ready. The 4 a.m. weather report was typically sketchy: "Strong westerly wind. Conditions oth-

In their holiday best, Newfoundlanders gather with picnic lunches to watch the final preparations and takeoff of Alcock and Brown's Vickers Vimy from the improvised airstrip known as Lester's Field.

erwise fairly favourable." It was the best forecast they would get. Brown dressed in his uniform, Alcock donned a blue serge suit and before dawn they drove to the field. Presently a boy on a bicycle set out after them; he was carrying sandwiches and a Thermos of coffee prepared for the trip by Agnes Dooley.

The wind slackened but remained gusty, and the airmen waited through the morning for it to subside. Special mail was loaded and seals were affixed to the fuselage.

Water for the radiators had been twice filtered,then boiled, to clear it completely of sediment. Alcock did not want to suffer a cooling system failure as Hawker had. The Vimy did not have a built-in lifeboat; its lifeboat was a detachable fuel tank. The compasses had been swung, Brown wrote, "more elastic shock absorbers were wrapped around the axles, and the navigating instruments were taken on board, with food and emergency supplies." Then, he added, "a large black cat, its tail held high in a comical curve," sauntered by. "Such a cheerful omen made me more than ever anxious to start." It also reminded Alcock to fetch their own mascots, two stuffed cats named Lucky Jim and Twinkletoe. Lucky Jim was lashed to a wing strut and Twinkletoe rode tucked into Brown's flying suit.

Coaxing their Vickers Vimy into the air above the raw landscape of Lester's Field near St. John's, Newfoundland, Alcock and Brown head toward Ireland. To reduce drag, Alcock had decided against adding a nose wheel, a feature that was designed to prevent the plane from plowing into the ground on landing.

There was one more problem: The wind shifted. In order to take off into the wind, the plane had to be pushed across the field; the takeoff would be slightly uphill. In the course of the move a fuel line was crushed. It took an hour to fix it, and Alcock and Brown used the occasion to eat lunch. "We'll have our next meal in Ireland," Alcock promised. A car drove up, klaxon hooting. It was the doctor who had prescribed the "tonic waters." He presented them with a bottle of whiskey, and Alcock took a jolt.

The wind continued to bluster, but the weather was as favorable as any the two fliers had seen during their weeks in Newfoundland. Alcock told Brown quietly that if they did not go now they probably never would. Brown agreed, and both men climbed into the cockpit. Alcock fired up the two engines. One after the other they threw out a cloud of blue smoke, then they settled into a smooth and steady beat. He raised his hand to the men who stood in front of the wings acting as brakes to hold back the plane. One of the men later remembered: "Alcock revved up the engines to full power. Then, on a signal, we all sat down on the ground and the plane shot forward."

Lurching from side to side in the wind, the Vimy lumbered more than 300 of its allotted 500 yards, "showing not the least desire to rise." Raynham watched in anguished silence. When they were "almost at the end of the ground tether allowed us," Brown recalled, the plane rose, skimmed narrowly over a stone dike and a row of trees, then sagged below a hill crest a few hundred yards away. The crowd, certain that the Vimy had crashed, started to run. The doctor pushed people aside, shouting, "Make way—they'll be needing me." But then the Vimy reappeared, airborne and climbing, with Brown waving from the cockpit. Brown looked at Alcock and "noticed that the perspiration of acute anxiety was running down his face."

Ships in St. John's harbor saluted them with whistle blasts they could not hear. At 1,200 feet, they crossed the coast. Flying under patchy clouds, they saw icebergs on the horizon. Brown radioed a message: "All well and started." Then the wind-driven generator propeller sheared off, leaving the radio powerless. There would be no asking for help or guidance.

Perhaps they would need none. The weather appeared favorable and they were riding a whistling tail wind. Brown now faced the formidable task of finding the way to Europe in a speeding aircraft. He had a standard naval sextant with which to shoot the sun or stars. The sextant had a crude artificial horizon in case clouds obscured the real horizon, as they so often did. Using a slide rule and navigation tables, he could convert the sextant readings into a position on the chart he carried open on his knees. When the sky clouded over, he could rely on dead reckoning—estimating the effect of wind velocity and the Vimy's air speed on the plane's position. The trouble with dead reckoning was that a crosswind could blow them as much as 50 miles off course in an hour, a potentially fatal deviation. To measure this sideward motion, Brown

Aids and comforts for the risky mission

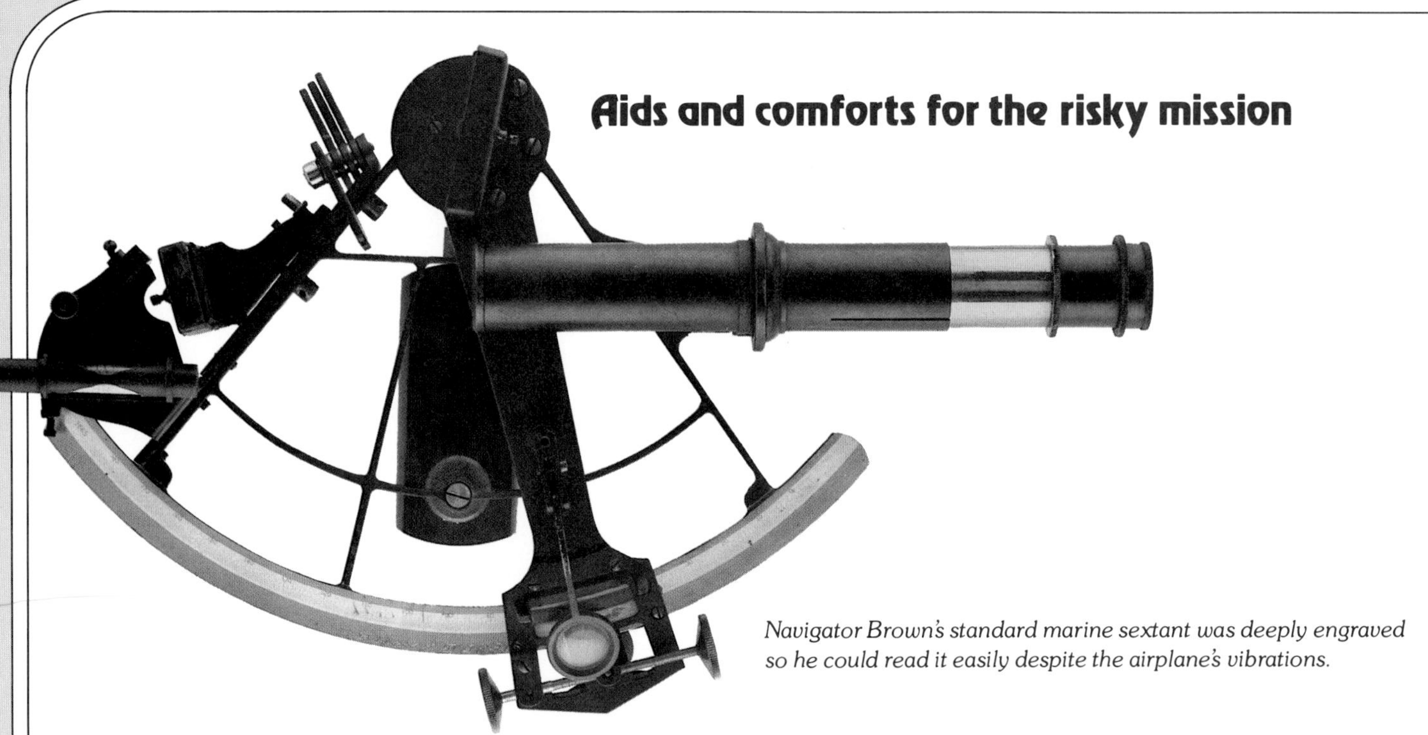

Navigator Brown's standard marine sextant was deeply engraved so he could read it easily despite the airplane's vibrations.

The Vimy's radio, whose interior is shown here, stopped transmitting when its wind-driven generator lost a propeller.

Alcock's blue serge waistcoat, worn under his flying suit, had an inflatable tube around the middle to act as a life belt.

Brown wore this electrically heated beige jacket under his flying suit. Wires passed through the fabric and to the soles of his boots.

Twinkletoe, a "daintily diminutive" stuffed cat, was Brown's mascot on the Atlantic flight.

had an instrument called a drift bearing plate. But he could use it only when he could see the water below; when clouds obscured the sea, crosswinds would be undetectable.

Kneeling in his seat, Brown pointed his sextant toward the southwest to make his first sun shots. Then, with drift and speed measured from the waves, he began to lay out their progress on the chart. He had barely acquired this basis for dead reckoning when, within an hour after takeoff, they passed over a vast fog bank and lost sight of the sea. Then the haze above thickened and blotted out the sun. There was no way to determine their position.

Half an hour later, the starboard engine made a noise like a machine gun firing at close quarters. "A chunk of exhaust pipe had split away," Brown wrote later, "and was quivering before the rush of air like a reed in an organ pipe. It became first red-, then white-hot, and softened by the heat, it gradually crumpled up. Finally, it was blown away." The exhaust from six of the engine's 12 cylinders ripped through the air a few feet from their heads. Flames once diverted by the exhaust pipe now played on a bracing wire. It glowed red-hot but did not break, and the flames did not appear to be reaching the flammable fabric skin of the aircraft. Alcock and Brown roared on, the brutal noise blasting their ears.

Next, the battery that warmed Brown's electric jacket failed, but he was not uncomfortable. The Vimy's cockpit, though open, was partially sheltered. At 7:30, Brown twisted around and dug out Agnes Dooley's sandwiches and opened a Thermos of coffee. He handed Alcock's share of the meal to him a bit at a time and Alcock ate with one hand on the control stick. Throughout the long flight, Alcock would never let go of the controls. After supper, Brown passed a scribbled note to Alcock: "I must see the stars."

Up they climbed until they broke out of the clouds into dim moonlight at 6,000 feet. The air at this altitude was much colder than they had experienced so far. With aching fingers, Brown turned the knobs of the sextant and found his position. Eight hours out, they had flown 850 miles, almost half the distance to Ireland. These figures translated into an average ground speed of 106 miles per hour, slightly more than they had anticipated. Alcock dropped to 4,000 feet and skimmed the top of the clouds, cruising along under a hazy moon. "An aura of unreality seemed to surround us as we flew onward towards the dawn and Ireland," Brown wrote. "The distorted ball of a moon, the weird half-light, the monstrous cloud shapes, the fog, the misty indefiniteness of space, the changeless drone, drone, drone of the motors."

At 3 a.m., as dawn was streaking the sky, a cloud appeared like a wall before them, and they plunged in. The plane careened in violent winds and Alcock, unable to see the horizon, became disoriented. "Left to its own devices," Brown recalled, the Vimy "swung, flew amok and began to perform circus tricks. Until we should see either the horizon or the sky or the sea and thus restore our sense of the horizontal, we could tell only by the instruments what was happening." But the Vimy's primitive blind-flying instruments were of no assistance. The bubble disappeared from the glass of the turn-and-bank indicator, the instrument that was supposed to tell the pilot whether the plane's wings were level. The air-speed indicator jammed at about 90 miles per hour, so that Alcock was not aware that he was flying the Vimy slower and slower, approaching the stall speed at which the plane would no longer be able to remain aloft.

Suddenly the Vimy shuddered, its flying speed exhausted. The nose dropped, and the plane whipped into a spin. Down they plunged through the blackness. The sinking altimeter told them they were falling; the revolving compass told them they were spinning. But "how and at what angle we were falling we knew not," reported Brown. "Alcock tried to centralize the controls but failed because we had lost all sense of what was central."

The altimeter registered 3,000 feet . . . 2,000 . . . 1,000 . . . 500. In the quiet of the throttled-back engines, Brown could hear the sea thundering below. He stuffed the flight log into his flying jacket, though he knew that if they hit the water there was almost no chance of survival in the icy embrace of the waves below.

Then, as abruptly as they had entered the cloud, they popped out of it. The ocean was less than 100 feet below but the plane was nearly upside down, so that the water appeared to threaten them from above. Alcock's ability as a pilot would never be more severely tried. At the sight of the water, his equilibrium returned. With only seconds to live, he snapped the plane out of the spin, leveled off and shoved the throttles forward. The propellers bit the air, and the plane flew along no more than 50 feet above the waiting embrace of the Atlantic. Brown felt that he could almost have reached out of the cockpit and touched the white-caps with his finger tips.

With the instruments registering again, Alcock started up into the murk, seeking the altitude that spells safety for pilots. Heavy rain fell, and as they gained altitude it turned to snow that coated the plane. Ice began to form. Several times Brown had to stand on his seat and, clutching a strut with one hand, use the other to clear the snow away from the face of a vital fuel gauge and from the tubes that governed the air-speed indicator. "The change from the sheltered warmth of the cockpit to the biting icy cold outside was startlingly unpleasant," he wrote. But they dared not descend. They had to climb above the clouds so that Brown could use his sextant again to pinpoint their position after flying aimlessly in the turbulence.

Somewhere above 11,000 feet, Brown caught a glimpse of the sun through the clouds and took the long-awaited sun shot. At that point, said Alcock, "we had no certain idea where we were," but Brown made some calculations and consulted his chart. They were a bare 80 miles away from Ireland.

Then a series of sounds like rifle shots cracked from the starboard engine. Alcock shut it down. Ice had partially covered its air intake, causing it to backfire. With the other engine idling, Alcock began a long and shallow glide toward warm air at a lower altitude. The plane could not fly on one engine; the question was whether the ice would melt before they struck the water. Down they went until they broke into the open 500 feet above the sea. Alcock restarted the dead engine, then advanced the throttles; the engines responded with a healthy roar. Twenty minutes later Alcock and Brown passed two islands, and 10 minutes after that they were over Ireland.

Alcock swung the Vimy along the coast, saw the radio masts of the military installation at Clifden and headed for what appeared to be a smooth green field nearby; it was actually an immense bog. Men on the ground tried to wave the Vimy away from its intended landing spot, but Alcock and Brown merely waved back, returning what they interpreted as a greeting. As the Vimy touched down on the marshy surface, its wheels dug in and the plane nosed over. Both men scrambled out, Brown nursing a bumped nose. They had been in the air for 16 hours and 28 minutes.

"What do you think of that for fancy navigation?" asked Brown. "Very good," said Alcock, and they shook hands.

The two men were soon enjoying a triumphal journey by train and ship to London. At every stop people thronged to greet aviation's

Nose buried in an Irish bog, the Vimy draws a small crowd of civilians and soldiers. Troops from a nearby military radio installation, the first men to arrive at the scene, thought that Alcock and Brown were joking when they claimed to have just arrived from North America.

newest heroes. "In Crewe station there was yet another crowd," reported the *Times* of London, "which insisted on the airmen leaving the train during the halt. Suddenly an Australian soldier called to a porter 'Up with him,' and Lieutenant Brown was lifted shoulder-high so that all the people could see and cheer him. Another soldier with assistance hoisted up Captain Alcock."

In London, the fliers rode to a reception at the Royal Aero Club, their open Rolls-Royce proudly flying the Union Jack. Lord Northcliffe, whose challenge had inspired their flight, was too ill to greet them or to award the £13,000 in prizes during a luncheon at the Savoy Hotel. In his stead he enlisted Winston Churchill, then Secretary of State for War and Air, to do the honor. On the following day Alcock and Brown were knighted by King George V.

Back in Newfoundland, Admiral Kerr decided there was no point in being second to Alcock's first and took the Handley Page instead to the United States, where it crashed in Cleveland, Ohio. The crew survived, but the bomber was demolished.

Britain's unabashed delight in the successful transatlantic flight soon turned to sorrow. In December 1919, Sir John Alcock, while delivering a new Vickers Viking to an air show in France, attempted to land in a fog, fouled a wing tip on a tree and crashed. He died a few hours later without regaining consciousness. Sir Arthur Whitten Brown, although he lived well into the 1940s, never flew again.

At London's Savoy Hotel, Winston Churchill presents Alcock and Brown with the Daily Mail prize for crossing the Atlantic. "I really do not know what we should admire the most in our guests," he said, as he gave them a check for £10,000, "their audacity or their good fortune."

First around the world

"After World War I, aviators were extremely restless," recalled aviator Leigh Wade, long after World War I. "Our commanders were restless, too, always looking for ways to show the strength of air power and gain public support." In pursuit of these goals, General Mason M. Patrick, who headed the Army Air Service after the War, announced what was at the time the ultimate test of plane and pilot: the first flight around the world.

Wade, a young lieutenant and pilot, eagerly volunteered in October 1923; six months later he and seven other Army airmen climbed into four seaplanes for the first leg of the 175-day United States World Cruise.

As the aircraft took off from Lake Washington near Seattle on April 6, 1924, the American aviators felt confident of success. Wade and the other three pilots, chosen from among the Army's best, had hand-picked their one-man crews (a single flier would serve each plane as both mechanic and copilot). The airplanes were Douglas World Cruisers, conceived especially for the flight by Donald Douglas, an up-and-coming young aircraft designer, and were named for American cities at the four main points of the compass: *Chicago, New Orleans, Boston* and *Seattle*. Each was a 400-horsepower single-engined biplane with pontoons that could be exchanged for wheels when necessary, depending on the landing sites along the way.

The Army had scheduled the flight and plotted the route to avoid the worst of the world's bad weather. Fuel dumps and supply depots were established to keep the planes on schedule and the Navy dispatched rescue ships to strategic locations in case a crew was forced down at sea.

For the men who flew the planes, the journey held all the allure of a magic-carpet ride to some of the most exotic points on the globe. It promised more than 26,000 miles of danger and excitement and, if the aviators succeeded, the acclaim of a world not yet awakened to the truly global potential of aviation.

Two Douglas World Cruisers are examined by townspeople at Clover Field, near Douglas Aircraft in Santa Monica, California. The planes were later flown to Seattle for the official start of the flight.

This map shows the route of the United States World Cruise with all 72 stops made by the two planes that completed the trip. From their starting point in Seattle, Washington, the aviators flew clockwise around the world to avoid four adverse weather systems: spring fogs in Alaska, typhoons in Japan and China, the monsoon season in Southeast Asia and winter storms in the North Atlantic.

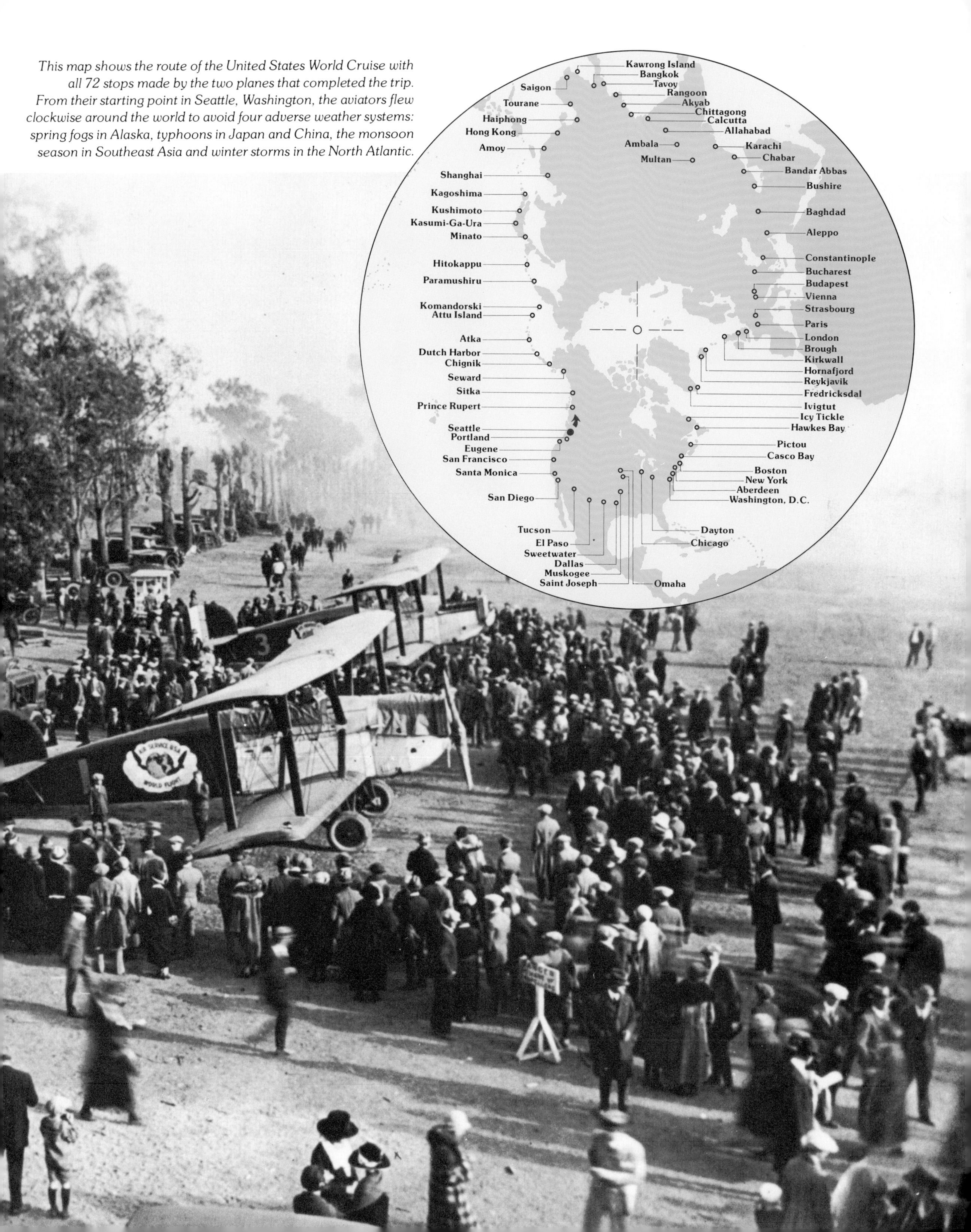

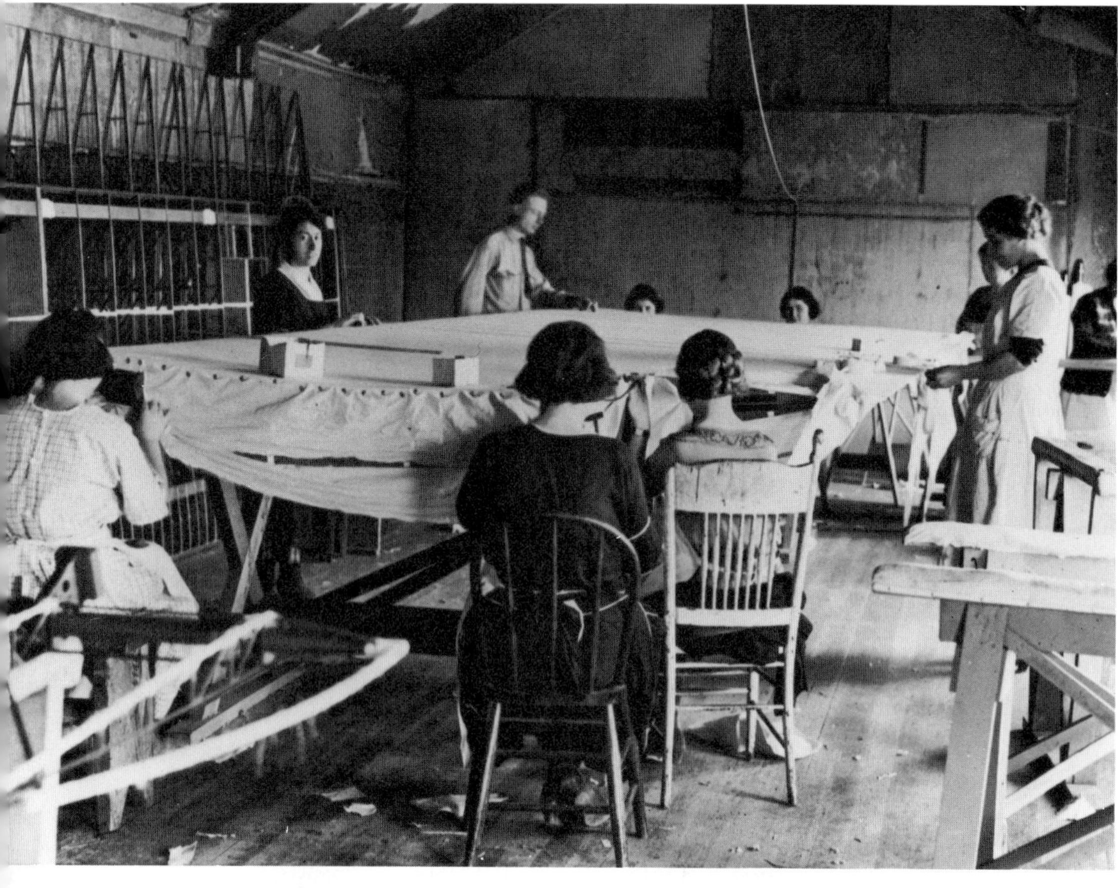

Seamstresses sew water-resistant linen around the wing of a World Cruiser in early 1924. Altogether, five aircraft were built, one prototype and four for the actual flight.

Six of the world fliers gather for a group portrait on the way to Seattle. They are, from left to right, Sergeant Henry H. Ogden, Lieutenant Leslie P. Arnold, Lieutenant Leigh Wade, Lieutenant Lowell H. Smith, Major Frederick L. Martin and Sergeant Alva L. Harvey. Of those pictured, only Arnold and Smith actually completed the full circuit. The other two finishers, Lieutenants Erik H. Nelson and John Harding Jr., had been delayed in San Diego for compass adjustments.

Lieutenants Nelson and Harding (right, center) stand aside as the New Orleans is christened with water from the Mississippi River. The Boston was splashed with water from Boston Harbor, the Seattle with water from Lake Washington and the Chicago with water from Lake Michigan.

Major Martin (foreground) pushes ice away from the Seattle, forced down in Alaska by engine trouble. The Seattle rejoined the flight after repairs but Martin flew into a fog-shrouded mountain several days later. He and his mechanic escaped serious injury, but the plane was demolished and command passed to Lieutenant Smith.

The three remaining World Cruisers prepare to land in the harbor at Kagoshima, Japan, on June 2, 1924, thirty-three days after the loss of the Seattle. "The shore was black with people," said Lieutenant Arnold of the crowd in which virtually every child waved an American or a Japanese flag. "It was an inspiring scene."

A World Cruiser hunts for a landing place on the boat-clogged Yangtze River near Shanghai. "We were amazed at the number of craft below us," recalled Lieutenant Nelson. "The river teemed with tens of thousands of junks, sampans and steamers."

Associated Press correspondent Linton Wells tries out a camel near one of the World Cruisers at Allahabad, India. In Calcutta, where the pontoons of the planes had been changed for wheels, Wells had wedged himself into the rear cockpit of the Boston; he was to help with aircraft maintenance since Lieutenant Smith was weak with dysentery. But four stops later in Karachi, Wells was evicted by Army order.

Lieutenant Smith smiles wearily as admiring French dignitaries welcome the American aviators to Paris on Bastille Day, July 14, 1924.

Photographed from an escort plane, the trio of World Cruisers wings toward London from Paris. From London, the world fliers went to Brough, England, where the planes were refitted with pontoons for the dangerous crossing of the North Atlantic.

After traveling more than 19,000 miles, the Boston sinks in the North Atlantic between the Orkney Islands and Iceland, the victim of an oil-pump failure. The Boston's crew, Lieutenant Wade and Sergeant Ogden, were rescued, but rough seas and equipment mishaps made it impossible to salvage the plane.

Ferried ashore by a launch from the U.S.S. Richmond, anchored on the horizon, Lowell Smith sets foot on North American soil at Icy Tickle, Labrador, on the 31st of August. "We landed after our travel," recalled Smith, "like the Pilgrim Fathers, on a large rock."

Flashing a rare smile, President Calvin Coolidge congratulates the airmen on September 9 on their arrival at Bolling Field in Washington, D.C. The President, who waited three hours in rain for the airmen, refused a suggestion to return to the White House. "Not on your life," said Coolidge. "I'll wait all day if necessary."

After a triumphant crossing of the United States, Lowell Smith is welcomed back to California by his parents, the Reverend and Mrs. Jasper Smith. "I'm proud and I don't care who knows it," said Jasper Smith. In Seattle 50,000 people welcomed the airmen at the official finish of man's first circumnavigation of the world by air.

2

Revving up for Paris

In late May of 1919, as Alcock and Brown fretted in Newfoundland to be off in quest of Lord Northcliffe's $50,000 prize, President Alan Hawley of the Aero Club of America in New York City received a letter mailed across town from the Hotel Lafayette. "Gentlemen," he read, "As a stimulus to courageous aviators, I desire to offer, through the auspices of the Aero Club of America, a prize of $25,000 to be awarded to the first aviator of any Allied country crossing the Atlantic in one flight, from Paris to New York or New York to Paris, all other details in your care. Sincerely, Raymond Orteig."

The name Orteig meant nothing to the president of the prestigious club. Who was this impetuous fellow who offered only half Northcliffe's prize money for flying nearly twice the distance?

Raymond Orteig had started life as a shepherd boy in France. Emigrating to New York at age 12, he worked as a bus boy, headwaiter and café manager and eventually acquired two hotels, the Brevoort and the Lafayette. His hotels exuded the atmosphere of Paris. During the War, French airmen assigned to temporary duty in the United States often stayed there, delighting Orteig with their stories of combat. Thus exposed to the excitement of flight, Orteig resolved to offer a prize as a way of keeping France in the forefront of world aviation.

The Aero Club took Orteig up on his offer. But for a long time the challenge drew no response, even though the immediate postwar years were brimful of exciting aeronautical events. Airlines established regular service across the English Channel and between major cities of Europe. By 1924, France and the United States, in lively competition, had exchanged possession of a number of flying records. In October 1922, General William "Billy" Mitchell of the United States flew 222.9 miles per hour to capture the world speed record from France, whose pilots had held it since the War. Four months later, Sadi Lecointe won it back for the French, then lost it again a year later to Lieutenant Russell A. Maughn of the United States, who flew at 236.5 miles per hour. In the competition to fly higher than anyone else, the French topped out in the fall of 1923 at 36,565 feet.

France for years kept the nonstop distance record to itself, but in 1922 an American, Lieutenant James Doolittle, flew 2,163 miles from Pablo Beach, Florida, to San Diego in less than 23 hours, stopping just once for fuel. In 1924, Lieutenant Maughn raced the sun from New York to San Francisco to prove that Army planes based anywhere in the United

A symbolic flier studies the North Atlantic on a globe in this illustration from the cover of a December 1919 issue of a French aviation magazine. Despite Alcock and Brown's momentous flight from Newfoundland to Ireland that summer, the broad expanse of ocean between Europe and North America remained a challenge to the world's aviators into the 1920s.

States could reach any of its borders in one day. Starting from Long Island at sunrise, Maughn touched down in San Francisco 17 hours later, one minute before the sun officially set. His average speed, including three stops for fuel, was 140 miles per hour.

In France the records were set mostly by civilians, but in the United States they were the work of the Army. Civilian aviation in America was represented chiefly by the fledgling airmail service and by the irrepressible barnstormers, gypsy showmen who came hopping over the horizon in war-surplus Jennies and de Havillands, sideslipped into pastures and offered rides to the locals at five dollars apiece.

But no one took aim at the Orteig Prize, because no plane yet developed was remotely capable of winning it. To be sure, Alcock and Brown had crossed the Atlantic in a Vickers Vimy. But remarkable as their flight was, it began and ended in remote places separated by the narrowest possible span of ocean. To double that distance, to fly from the largest city in North America to the capital of France: That was Orteig's challenge and it whetted the appetites of the world's best fliers.

Seven years passed before the first candidate boldly revved up his engines for a go at the prize. He was René Fonck of France, the leading Allied fighter pilot of World War I, with 75 confirmed victories. Fonck's fellow aviators considered him cold and haughty, but they respected him nevertheless as a superior pilot.

In search of an airplane that could cross the Atlantic, Fonck sailed to the United States early in 1925. He soon was chosen by an American syndicate to pilot an aircraft that had already been ordered from Igor Sikorsky, a Russian designer who had fled the Revolution of 1917. With the financial backing of fellow exile Sergei Rachmaninoff, the famed pianist and composer, Sikorsky had reestablished himself in a leaky hangar at Roosevelt Field on Long Island. There he built the S-35, a large biplane that had three 425-horsepower air-cooled radial engines

The Sikorsky S-35 stands outside the hangar where it was built at Roosevelt Field, New York, in September 1926, awaiting reports of favorable weather over the North Atlantic before embarking for Paris. Arrayed in front of the giant biplane are the dozens of 50-gallon drums of gasoline needed to fill the tanks in its upper wing and behind the two outboard engines. All told, the S-35 was to take off loaded with 15,000 pounds of fuel and oil to supply its three 425-hp Jupiter engines.

made in France. Built to be used eventually as a passenger airliner, the S-35 had a cabin 15 feet long, trimmed in mahogany and leather and furnished with wicker seats and a divan. Lightly loaded, the plane flew almost perfectly, even with one engine shut down. With the necessary 2,380 gallons of gasoline aboard for the Atlantic crossing, however, it would weigh 28,860 pounds—not an impossible load, but some 10,000 pounds more than the plane was designed to lift. Sikorsky scheduled a series of tests to prove that the S-35 would fly when fully fueled, but Fonck, pressured by news that others were about to try for the Orteig Prize, canceled the tests to save time.

On September 20, 1926, mechanics fueled the S-35. Fonck climbed aboard with a cosmopolitan crew of three: Lieutenant Lawrence W. Curtin of the United States Navy, the copilot and navigator, sat next to Fonck in the cockpit; Charles Clavier, a young French radio operator, and Jacob Islamov, a Russian mechanic, sat behind them, inside the cabin. At the last moment a *bon voyage* gift arrived—it was a bag of *croissants,* presumably from Orteig. Fonck hefted it as if considering its weight, then tossed it inside.

The big plane started down Roosevelt Field's soft, mile-long runway, which ended on a bluff just above a somewhat smaller airfield named Curtiss Field. The S-35 rocked and trundled along, accelerating slowly. It struck a ridge in the runway and a wheel flew off the auxiliary landing gear that had been added to help support the plane's weight. The plane slued to one side; Fonck corrected. By now he had no choice but to continue the takeoff run. Like almost all planes of the day, the S-35 had no brakes, and there was no way to stop before reaching the end of the runway. Fonck hoped that he could build at least enough speed to float to a landing on Curtiss Field below, but at the runway's end the plane had not even attained gliding speed. It cartwheeled into a gully between the two fields and erupted in flame and smoke. Fonck and Curtin

Designer Igor Sikorsky (at left, below) and chief pilot René Fonck, France's leading World War I ace, wave triumphantly from the S-35's cockpit on the occasion of the plane's first test flight. Sikorsky, a Russian émigré, had been designing highly successful multiengined planes since 1912, and the S-35 was no exception; it performed brilliantly in early tests. But when the overloaded craft attempted to take off for Paris, it failed to clear the runway, crashed and burned (lower right).

scrambled out of the cockpit, but Clavier and Islamov were trapped in the cabin and burned to death.

"It is the fortune of the air," was Fonck's fatalistic observation. Almost immediately Sikorsky began preparations to build a new plane and Fonck planned to fly it.

The crash emphasized one of the dangers of a flight across the full breadth of the Atlantic, but that did not mean such a flight was impossible. The whole flight would be hazardous: Navigation error, engine failure or storms at sea could end the adventure disastrously. But the initial hurdle was getting off the ground; any plane carrying enough gasoline to fly the 3,600 miles between New York and Paris would be taking off dangerously overweight.

What made such a flight appear possible at all were the enormous strides that aeronautical engineers were making to create more efficient aircraft, capable of carrying greater payloads and staying in the air longer. The latest air-cooled engines were far more reliable than older air- or water-cooled types and could produce at least the same horsepower while weighing about 30 per cent less than earlier engines. Light but strong metals began to replace much of the heavier wood used in aircraft construction. Wing surfaces were being shaped more effectively to create more lift, and the monoplane, an aircraft with a single wing, was gaining favor over the biplane. After years of debate, most aeronautical engineers—though Sikorsky was not one of them—had come to agree that one wing caused less drag than two and could lift more, partly because a second wing interfered with the flow of air around the first.

One aircraft more than any other of the age represented the ideal marriage of new engine and airframe technologies. In 1926 the Wright Aeronautical Corporation of Paterson, New Jersey—which retained the Wright brothers' name, even though Orville, the surviving brother, no longer was associated with it—had perfected the J-5C Whirlwind engine *(box, page 54-55)* and the company decided to build a plane of its own to show off the latest Whirlwind to best advantage.

One of Wright's test pilots, a racer and barnstormer named Clarence Chamberlin, had once entertained crowds in a biplane designed by Giuseppe Bellanca, a soft-spoken Sicilian who had come to America before the War. Bellanca's plane, Chamberlin remembered, "did a lot more things than other barnstorming planes, which were quite clumsy by comparison." He recommended Bellanca to the Wright company, and Wright hired Bellanca to build the perfect plane for its engine.

The Wright-Bellanca was a single-engined, high-winged monoplane that weighed a mere 1,850 pounds when empty. The fuselage as well as the four distinctively broad struts supporting the wing were shaped like airfoils for additional lift. With the 200-horsepower Whirlwind engine, the plane could carry one and one third times its own weight and cruise at 100 miles per hour, going farther on a gallon of gasoline than any other aircraft of the day. From the moment of its unveiling, it was generally agreed that no better airplane existed in the world.

The Whirlwind's nine cylinders surround its crankcase.

An advertisement trumpets the engine's popularity.

The engine that made it possible

In a period of six weeks in 1927, three airplanes flew from New York to the European mainland—something no plane had ever done before. All three were powered by the same engine, the Wright Whirlwind J-5C.

This remarkable power plant was the latest in a series of 200-horsepower air-cooled radial engines initially designed by a former racing-car engineer named Charles L. Lawrance. Perfected during seven years of testing and in millions of miles of service, the Whirlwind was, by 1927, one of the most reliable aircraft engines in the world.

Like other air-cooled engines, it required no complex liquid cooling system, failures of which caused as many as a third of the forced landings by aircraft with contemporary liquid-cooled engines. Moreover, elimination of the cumbersome cooling apparatus made the Wright Whirlwind as much as 25 per cent lighter than liquid-cooled engines of comparable power.

Lightness and reliability, though critically important for long-distance flights, were not enough. Fuel economy was also essential. Earlier air-cooled engines had guzzled a profligately rich but cool-burning air-fuel mixture to keep from overheating. But the J-5C had completely redesigned cylinders that dissipated heat so effectively that the engine could burn a leaner, more economical mixture without overheating.

The significance of the Whirlwind was acknowledged in the awarding of the Collier Trophy, America's most respected aviation award, for 1927. The prize in that banner year went not to the pilots who had flown the Atlantic or to the builders of their planes but to Charles L. Lawrance, the designer of the engine that had made it all possible.

A week or so after René Fonck's dramatic failure in September of 1926, an obscure airmail pilot, Charles Augustus Lindbergh by name, was flying across Illinois in a war-surplus de Havilland, bearing a half-empty mail sack to Chicago. As he flew, he daydreamed. How wonderful it would be to sail ever onward. With enough fuel, a plane would be a magic carpet, able to carry him anywhere.

"In a Bellanca filled with fuel tanks I could fly on all night, like the moon," he thought. "How far *could* it go if it carried nothing but gasoline? With the engine throttled down it could stay aloft for days."

Even as a boy in rural Minnesota, Lindbergh had dreamed of flying, but whenever an itinerant barnstormer landed in a local field or an air show stopped at a town nearby, he remained a mere spectator. His mother never let him go up for a ride. Lindbergh's fascination with flight represented more than boyish enthusiasm. The airplane symbolized both his personal sense of remoteness from the world around him and the keen interest in machinery he had developed at an early age. A mechanic who repaired equipment at the Lindbergh farm was astounded to find that nine-year-old Charles knew as much as he did about gasoline engines. At age 11, almost as soon as his legs were long enough to reach the pedals, Charles took over the family automobile as chauffeur and mechanic. Danger seemed to draw him; on an old motorcycle, he performed feats that amazed his friends. He began keeping a self-improvement chart that listed 65 admirable traits—altruism, ambition, brevity, tact and unselfishness among them—on which he scored himself daily. As he grew older, he did not smoke, drink or gamble, and he generally avoided women and other pleasurable pursuits.

Despite such discipline, Charles was an unspectacular student; school bored him. But to please his parents he agreed to go to college and in 1920 he entered the University of Wisconsin to pursue a curriculum in mechanical engineering. After he had spent 17 dismal months there, his grades were so low that the school asked him to leave.

One of Lindbergh's friends at the university, a flying enthusiast himself, had acquired a brochure from the Nebraska Aircraft Corporation, which advertised a flying school at its plant in Lincoln. Lindbergh, for whom the step from idea to action was a short one, immediately set out for Nebraska, although, as he put it years later, "I'd never been near enough to a plane to touch it before." From the moment he entered the hangar at Lincoln, the aroma alone convinced him that he had made the right decision. "I can still smell the odor of dope that permeated each breath, like ether in a hospital's corridors," he wrote. He had found his vocation—the sky and the life that permitted him to fly.

Lindbergh went up for the first time a week after he arrived in Lincoln, and he knew that his instinctive embrace of flying had been correct: It was glorious. He paid the Nebraska Aircraft Corporation's $500 tuition, for which the company guaranteed him all the mechanical training necessary to maintain and repair a plane—a pilot in 1922 had to be his own mechanic—plus 10 hours of flight instruction leading to a solo flight and a flying certificate. Lindbergh, it turned out, was the school's sole student, and though the company employed pilots and mechanics who could show him the ropes, it had no organized training program. In fact, the pilot who was being paid extra to teach him was reluctant to do so—or even to fly himself—because he had recently seen a friend die in a crash. It took six frustrating weeks for Lindbergh to accumulate a mere eight hours in the air.

Not that he wasted the intervening days. He learned how to keep an engine running and how to repair it when it stopped. He mastered the technique of fitting a new propeller to the engine shaft and he practiced the lock stitch used to mend or patch a plane's fabric.

Lindbergh had not yet soloed when his instructor left town to take another job. Before departing, however, he certified to Ray Page, president of the company, that Charles—or Slim, as he was called—was qualified to fly alone. But since Lindbergh did not have an additional $500 to post against possible damage to the plane, Page refused him permission to solo. Shortly thereafter, Ray Page sold the last plane in his inventory. Lindbergh, seeing the opportunity to solo about to fly out of his life, quickly raised $100, including a partial refund of his tuition and several weeks' back pay that Page owed him for sweeping out the hangar, to buy a parachute. Some weeks later in Bird City, Kansas, he teamed up with a pilot named H. J. "Cupid" Lynch for a season of barnstorming across the prairies in a Lincoln Standard biplane. Lindbergh walked the wings as they approached each new town to advertise the thrills to come, and made trick parachute jumps that awed the crowd, but Lynch never did let him fly the plane alone.

To pass this milestone, Lindbergh finally purchased his own wings—a war surplus Curtiss JN-4 trainer, known as the Jenny. The plane, with a wingspan of nearly 45 feet, and in Lindbergh's model, an engine of only 90 horsepower, cost him $500, complete with a fresh coat of olive-drab dope. Lindbergh took delivery in Georgia, and though he knew that the Jenny handled much differently from the planes he had trained in, he asked for no pointers, too embarrassed to admit that he had never flown alone. It was his plane and he would fly it himself.

First, however, he decided to taxi for practice. Years later, Lindbergh described what happened next: "I opened the throttle—cautiously. The Jenny swerved a little. I kicked opposite rudder. It swerved the other way. I straightened out—opened throttle more—the tail lifted—a bit too high—I pulled back on the stick—the tail skid touched—I pushed forward—pulled back—before I knew it I was in the air! I cut the throttle—dropped too fast—opened it wide—ballooned up, right wing low—closed the throttle—yanked back on the stick—bounced down on wheel and wing skid." Fortunately, the hard landing did not break anything; by evening—with a little help tactfully volunteered by a concerned bystander—the young pilot watched the sun set from 4,500 feet, alone in his own plane.

A week later he was off on a one-man barnstorming tour which

Years that shaped a future hero

Writing late in his life, Charles Lindbergh reflected that his early years had been influenced by three environments: his family's farm in Minnesota; the Detroit laboratory of his grandfather, an eminent dentist; and the city of Washington, where his father spent 10 years in Congress and where Charles went to school.

He loved his adventurous outdoor life at the farm, where he could hunt in the woods or swim in the nearby Mississippi, and his natural mechanical curiosity was aroused by the technical wonders of the laboratory. But he was restless in the city, and he disliked school. He enrolled reluctantly in the University of Wisconsin but left in his second year to try an occupation that promised to satisfy his fascination with machines and his yearning for adventure: He became an aviator.

Lindbergh's mother, Evangeline, a chemistry teacher, was devoted to her only child.

Lindbergh's father campaigned successfully for Congress from Minnesota in 1906. His own father had served in Sweden's parliament.

Young Charles shoots at targets behind his grandfather's house in Detroit. He had two guns by the time he was seven and often went bird hunting with his father.

His grandfather's laboratory (left) was a magical place for young Lindbergh. Here the ingenious dentist introduced the boy to the machines he employed to contrive not only novel dental apparatus but a range of inventions from furnaces to baby rockers.

Charles stands beside his Excelsior motorcycle in 1921. Lindbergh rode the cycle on trips as far as Florida, enjoying the sensation that the machine was an extension of himself—a feeling he would have again while flying airplanes.

A beaming smile belies Lindbergh's unease as a University of Wisconsin student. He later blamed his lackadaisical study habits on his never having spent a full year at any school as his family shuttled between Minnesota, Detroit and Washington.

Slim Lindbergh (left) and his friend Bud Gurney lean against a Lincoln Standard, the type of plane in which Lindbergh first flew in 1922.

began, by chance, in Maben, Mississippi, when he was forced down there in the dark by a storm. Low on cash, Lindbergh offered rides at five dollars apiece. Privately, it amused him that he was being paid to carry passengers on flights during which he was still teaching himself to fly. But some sympathy must be reserved for his unwitting customers, like the man who cowered, terrified, in the front cockpit while the Jenny, by means of shuddering stalls and roller-coaster descents, proved to Lindbergh that it was too underpowered to fly a loop.

Lindbergh found that he fitted happily into the gypsy life of a barnstormer. He liked being on his own. He slept under the upper wing in a hammock rigged between struts and he traveled with little more than a toothbrush, a razor, soap and a clean shirt. He learned quickly to size up a pasture from the air for rocks, stumps, ditches or fences; he learned to be wary of cows, which liked the taste of airplane fabric and had been known to eat a plane to bare ribs in a few minutes. Having landed, he would send to town for a can of gasoline and await passengers. Some of them were a little eccentric, too. One of Lindbergh's favorites was the fellow who fired a pistol from the air and later remarked with pleasure that he had "shot this town up afoot an' a' hossback," and now he had done it "from a airplane."

The life of a barnstormer was dangerous. In 1923 alone these daredevils were involved in 179 serious accidents, in which 85 were killed and another 162 injured. Lindbergh once watched as the wing of a friend's more powerful Jenny fell off during a loop. The man's foot became wedged between the fuselage and a wire brace as he tried to parachute to safety. Lindbergh later lifted his body from the wreckage.

Lindbergh himself once landed in the town square of Camp Wood, Texas, and found that on takeoff, because the wind had shifted, he would have to thread his plane between light poles only two or three feet farther apart than the span of his wings. He did not make it. A wing tip struck one of the poles, and the plane spun around and crashed into a hardware store, damaging some crockery but not the pilot. Landing in Kansas on another occasion, Lindbergh ground-looped—spun around on the ground after hitting a stone. He did the same thing in Minnesota, this time after landing in a swamp.

Lindbergh gave up barnstorming in 1924, not so much because it was hazardous but because of its unpromising future. The profession, he decided, had become overpopulated, the countryside overbarnstormed; a pilot could barely make expenses. Moreover, Lindbergh realized that to get ahead in aviation, he would need better credentials than could be earned giving goose bumps to passengers at five dollars a ride. So he sold his Jenny and joined the outfit that offered the best flying instruction then available: the United States Army Air Service.

The Army sent him to flying school at Brooks Field near San Antonio, Texas. Not surprisingly, he was soon flying circles around his classmates, most of whom never had flown before. Lindbergh, blessed with the quick reflexes of an athlete and with 300 hours of seat-of-the-pants

SOUTHERN ILLINOIS HARD ROAD BASEBALL ASSOCIATION CARTERVILLE TEAM PRESENTS

Vera May Dunlap's

Flying Circus

In conjunction with their opening game

Sat. & Sun., May 9 & 10

At Carterville Base Ball and Aviation Field

1 MILE SO. OF CARTERVILLE AT CROSS ROADS

Carries Mills vs. Carterville

By special arrangements Vera May Dunlap will appear IN PERSON and will positively stand erect on top of the upper wing of the airplane without any visible means of support whatsoever while her pilot loops the loop, defying all laws of gravitation. Miss Dunlap carries with her a fleet of airplanes under the personal direction of Capt. Frank T. Dunn, the Canadian Ace. The only flier who has successfully looped the bridge on a navigable stream. Also Herbert Budd in the swing of death. T. Gurney the fastest wing walker and aerial performer in the world. Enslow, Hissel, Mann, Armstrong and Brown, pilots. Last but not least is Lt. Chas. A. Lindberg who saved his life by jumping from an aerial collision on March 6, 1925 in Kelly Field, Texas. He is now in the U. S. Air Service and will positively be in Carterville May 9-10. Herbert Budd will change from the top of one plane to another without any rope ladder. Admission to the ball game, including the airplane circus will be

50 cents

The gates will be open at 10:00 A. M. Parking space for cars free. Miss Dunlap also has three passenger carrying planes with licensed pilots for those wishing to ride. Price to ride **$3.00.** Look for airplane distributing programs as some will contain free riding and admission tickets.

Carterville Herald Print.

Buried in this broadside for Vera May Dunlap's Flying Circus is the misspelled name of Charles A. Lindbergh, who was picking up odd barnstorming jobs in 1925. The collision mentioned in the poster happened during Lindbergh's military flight training in Texas earlier that year.

barnstorming behind him, had become a skilled pilot. Army training would add the polish needed to make him a finished professional.

Yet, for all the discipline the Army imposed, flying at Brooks Field on occasion was every bit as dangerous as barnstorming. During advanced training, Lindbergh took turns with his classmates diving on a target plane in simulated combat. Once, as he broke off his mock attack, a sudden jolt smacked his head against the cockpit cowling. He looked up to see that another plane had crashed into him and that the student pilot was preparing to jump.

"Our ships were locked together with the fuselages approximately parallel," wrote Lindbergh in his official report on the incident. "My right wing was damaged and had folded back slightly, covering the forward right-hand corner of the cockpit. Then the ships started to mill around and the wires began whistling. The right wing commenced vibrating and striking my head. I climbed out past the trailing edge of the damaged wing and, with my feet on the cowling on the right side of the cockpit, which was then in a nearly vertical position, I jumped backwards. The wreckage was falling nearly straight down and for some time I fell in line with its path and only slightly to one side. Fearing the wreckage might fall on me, I did not pull the rip cord until I dropped several hundred feet into the clouds." Lindbergh and the other pilot were lucky to have survived; their class was the first to which the Army had issued parachutes.

Lindbergh spent only half his time at Brooks Field in the cockpit. The rest he spent in what for him had always been an alien arena: the classroom. His first grade was a lowly—and frightening—72. It was barely passing; one misstep and he could wash out of the school. "I concluded," he wrote many years later, "that the surest way of passing all seventy-plus examinations would be to strive for the highest marks I could get. I began studying as I had never studied before—evenings, weekends, sometimes in the washroom after bed check." The effort paid off; Lindbergh graduated in March 1925 at the top of his class.

The peacetime Army was short on funds for aviation, and Lindbergh, having won his wings, was released from active duty. That summer he joined a Missouri National Guard squadron based at Lambert Field, St. Louis, and hired on as chief pilot for the Robertson Aircraft Corporation there. His assignment would be to fly an airmail route between St. Louis and Chicago that the company expected to be awarded by the United States Postal Service.

Lindbergh laid out the 285-mile route, choosing landing fields and making arrangements with local postmasters and fuel suppliers. The fields were no more than small pastures—the one at Springfield, Illinois, occupied less than 40 acres. The Robertson brothers, William and Frank, who owned the company, could not afford beacons or radio; at most fields, the only lights were lanterns set out by mail-truck drivers.

The chief hazard in flying the mail lay in the regularity that it demanded. The Robertson brothers' route was a trunk line that had to connect with the transcontinental mail at Chicago. The schedule required flying after dark, no matter what the weather, in conditions that kept other planes grounded. Weather reporting was uncertain and Lindbergh, once he began flying the route, usually ignored the forecasts. If he could see to take off, he did so. Time and again he ghosted along a hundred feet or so under a sagging cloud, peering through rain squalls for the next landmark, with one eye cocked for the smoothest field should the clouds close in and force him down. At night he was guided by the lights of familiar towns and farmhouses. One farm boy along the route rigged a 100-watt light in his yard and wrote to the grateful pilot, "Maybe it will help. I will keep it lit every night."

On two occasions fog and snow obscured the ground so completely that Lindbergh ran out of gas before he could find a place to land and had to abandon his plane in mid-air. Once he climbed to 14,000 feet before he jumped. "If I could see the stars," he remembered thinking, "I wouldn't mind so much diving out into the storm." At last, looking up "from the bottom of a giant funnel in the clouds," he caught the faintest glimpse of starlight. Then he bailed out into a storm so opaque that he

Charles Lindbergh loads the first sack of mail aboard a Robertson Aircraft Corporation D.H.4 for the inaugural run of the St. Louis-to-Chicago airmail service in April 1926. By then a veteran barnstormer and a lieutenant in the Army Air Service Reserve, Lindbergh, as Robertson's chief pilot, was responsible for laying out the airmail route and making the first flight.

saw the ground only an instant before he struck it; the plane crashed a mile away. Next morning, Lindbergh retrieved the mail, some of it soaked with oil, and delivered it to Chicago in another plane.

Though foul weather was full of danger, Lindbergh preferred it to clear skies. In good weather he found nothing to pit himself against. "You simply sit," he wrote, "touching stick and rudder lightly, dreaming of the earth below, of experiences past, of adventures that may come." It was on such an evening, over Illinois, that Lindbergh's thoughts wandered to his mail plane and the modest range that required it to land too soon, too often. His mind conjured up the grand possibilities that would be open to him—if only he owned a Wright-Bellanca.

"It's fast, too. Judging from the accounts I've read, it's the most efficient plane ever built. It could break the world's endurance record and set a dozen marks for range and speed and weight. Possibly," Lindbergh remembered thinking in late September, 1926, "I could fly nonstop between New York and Paris."

The idea took hold, and immediately he began to plan. He would need perhaps $10,000 to purchase the Bellanca. He had $2,000 of his own, carefully saved; perhaps he could raise the rest among St. Louis businessmen. A St. Louis-sponsored flight would show that the city on the Mississippi was an up-and-coming place. And the Orteig Prize, if he won it, would amply repay the investment.

Weight, Lindbergh knew, was a major problem. Fonck's crash had proved that. The Frenchman had seemed almost cavalier about weight, with that unnecessarily luxurious cabin in his plane, not to mention the bag of *croissants*. The Sikorsky had three engines. If one failed, it could limp back home on two—perhaps. But three engines, Lindbergh reasoned, meant that three times as many things could go wrong; in particular, if one quit on takeoff, they tripled the chance of failure. No, it had to be a small plane, a simple plane, one with every ounce of weight considered—and flown by a single man. So far, all the Atlantic attempts had been at least two-man endeavors, sometimes more. But were they all necessary? If Lindbergh could navigate for himself, a second man might not be worth his weight in fuel.

Back in St. Louis, Lindbergh set his plans in motion. He persuaded the Robertson brothers to give him time off to pursue the project and to let him use the company name. He persuaded a businessman who was taking flying lessons to help finance the flight; others would follow. Next he began looking for a suitable airplane. He sounded out a salesman for the Fokker company, which had produced great fighting machines for Germany during the War and afterward had opened a plant in New Jersey. Yes, Fokker had plans for a plane with transatlantic range; it would have three engines and would cost almost $100,000. Lindbergh blanched. He had in mind a single-engined aircraft. Sorry, sniffed the salesman, Fokker would not consider selling a single-engined airplane for such a dangerous flight.

The Wright-Bellanca was the plane that Lindbergh really wanted. The problem was how to get the Wright company to treat him more seriously than Fokker's salesman had. The first step was to place a long-distance telephone call to the Wright plant in Paterson, New Jersey. That minor extravagance opened the door to Lindbergh. A short time later he visited the company and met the designer, Giuseppe Bellanca, who assured Lindbergh that the plane could cross the Atlantic. In contrast to Bellanca's unreserved enthusiasm, the Wright company was reluctant to turn the plane—in which it had invested so much money and prestige—over to an unheralded Midwesterner for a solo mission that was very likely to be suicidal. If Lindbergh failed, the publicity would be devastating. So the company put Lindbergh off—and ultimately refused to sell him the plane.

Slim Lindbergh was by no means the only pilot planning to fly the Atlantic in 1926; in fact, he trailed far behind many others, nearly any of whom seemed a better bet than he. In Europe alone, aviation enthusiasts could count at least a dozen efforts; the most advanced among them was that of Lieutenant Charles Eugène Jules Marie Nungesser, France's third-ranking ace of World War I. He planned to cross the Atlantic flying a single-engined biplane west from Paris to New York, a risky venture that Nungesser calculated could succeed only if he faced no head winds for the first hours of the flight.

In the United States, Commander Richard E. Byrd had purchased a Fokker Trimotor for the flight. Byrd had chosen the same type of plane for his recent adventures in the Arctic. In 1926 he had become world famous for an aerial expedition on which he claimed—though he was seriously disputed many years later—to have flown over the North Pole. The Fokker had been built as a single-engined monoplane and had been transformed into a trimotor by slinging an additional engine under each wing. Money to buy the $100,000 airplane came from Rodman Wanamaker, an owner of department stores in Philadelphia and New York. Byrd, who had attended flight school after graduating from the Naval Academy in 1912, was a competent but not a superior pilot. Realizing this, he concentrated on organizing aerial epics and navigating. He usually left the flying to such crack pilots as Floyd Bennett, who had been with Byrd on the North Pole excursion and would pilot the new Fokker, christened *America,* on the transatlantic effort.

A second Naval Academy man, Lieutenant Commander Noel Davis, also had a plane being readied for the Atlantic. Like Byrd, Davis excelled as a navigator rather than as a pilot. He was also a student of aerodynamics who saw the Atlantic flight as a technical problem that Americans, as a matter of pride, should be the first to solve. The American Legion, an association of war veterans, agreed. To boost its convention in Paris, scheduled for 1927, the Legion put up $100,000 to support Davis' entry, a special version of the Keystone Aircraft Corporation's Pathfinder bomber, which would be named *American Legion.*

An irrepressible flying princess

In early 1927 two British aviators entered the great transatlantic race, making no mention of any additional members of their flight crew. But when their single-engined Fokker, the *St. Raphael,* finally set out from England on August 31 bound for Ottawa, it carried a third aviator whose identity, once revealed, caused an international sensation.

Princess Anne Lowenstein-Wertheim, the widow of a German prince, was an irrepressible 62-year-old nonconformist who had been born Lady Anne Savile in England. An athlete and a feminist, she had for years literally flown in the face of her staid family's opposition to her escapades. In 1912 she became one of the first women to fly across the English Channel, and later she took part in pioneer long-distance flights ranging as far as Egypt. Deciding to become the first woman to fly the Atlantic, she hired two well-known pilots, Leslie Hamilton and Fred Minchin, and bought the *St. Raphael*—all in strict secrecy. But the news leaked out, and the Princess went into hiding to avoid her outraged family.

Moments before takeoff, she sped up the runway in a friend's car. "It is a great adventure," she calmly told the press, then boarded the plane. The *St. Raphael* was spotted once by a ship at mid-Atlantic—and was never seen again.

Princess Lowenstein looks regal in an early photograph.

Thus, by the late fall of 1926, a number of well-financed attempts at a transatlantic flight were under way. Charles Lindbergh's, however, was not among them. He had succeeded in putting together a supporting organization in St. Louis headed by a broker and a banker. But he had not officially entered the contest because he had no plane. After deciding against selling its Bellanca to Lindbergh, the Wright aircraft company sold it instead to a young New York businessman named Charles A. Levine. Levine came from Brooklyn; he had branched out from his father's scrap-metal business into selling automobiles, making himself rich. When he heard that the Bellanca was for sale, he decided to buy it and manufacture more like it. He formed the Columbia Aircraft Company with offices in Manhattan's Woolworth Tower, then the world's tallest building. He named the plane *Columbia.* Giuseppe Bellanca, the designer, and Clarence Chamberlin, the pilot who had introduced Bellanca to the Wright company, came along as part of the sale. Both believed that the Bellanca needed only the addition of extra fuel tanks to be made ready to cross the Atlantic. And Levine appreciated how helpful to his new enterprise a successful flight would be.

Lindbergh, meanwhile, continued to search. Though the aviation industry was lean and hungry for work, the companies he approached were unwilling to take part in a project that seemed certain to fail. In near-desperation, he turned to an obscure California company named Ryan Airlines. The Ryan plant occupied two floors in a wing of a fish cannery on the San Diego waterfront, where the smell of fish blended with the banana-like odor of airplane dope. The company produced a high-winged monoplane, the Ryan M-2, that had performed admirably on West Coast mail routes. In many ways the M-2 resembled the Wright-Bellanca; and little wonder, for T. Claude Ryan, the company's designer and founder, had included many of the same features in the M-2 that had made the Bellanca such a success. Using Robertson's name, Lindbergh telegraphed Ryan: CAN YOU CONSTRUCT WHIRLWIND ENGINE PLANE CAPABLE FLYING NONSTOP BETWEEN NEW YORK AND PARIS STOP IF SO PLEASE STATE COST AND DELIVERY DATE.

Claude Ryan knew the Robertsons and considered the wire something of a godsend. Recent business disagreements had prompted him to sell his share of the company to his partner, Franklin Mahoney, although Ryan had continued on as general manager. Lindbergh's order would help ease a severe cash shortage, and the flight, if it succeeded, would undoubtedly bring in more orders for airplanes. Before he responded, Ryan consulted Donald Hall, the company's first and only aeronautical engineer, who had been hired just three weeks earlier. The company already had a design for an enlarged version of the M-2, which would carry five passengers. With some changes, exchanging passenger space for fuel capacity, Hall and Ryan decided it just might be capable of reaching Europe. Ryan telegraphed an offer: $6,000 plus the cost of the engine, delivery in three months. By return wire Lindbergh asked, could it be delivered any sooner? Two months, Ryan answered.

Several days later, the Ryan company received another telegram: OUR PILOT CHARLES LINDBERGH WILL ARRIVE SAN DIEGO MONDAY ROBERTSON AIRCRAFT.

On February 23, 1927, Lindbergh arrived in San Diego by train; he sensed that he carried with him a tremendous responsibility. Only Ryan would build him a plane, but Ryan had never built a plane like the one he needed. If he placed the order, his capital would be exhausted, and there was no guarantee that the new plane would have the required range. But time was critical; every day brought his competitors closer to readiness. Lindbergh was on his own. His first step was to go up for a flight in one of Ryan's M-2 aircraft, which he found more than satisfactory. He would have to base his decision on his estimate of Don Hall's capacity to make the necessary changes in design.

From their first meeting, Lindbergh also found Hall more than satisfactory. The 29-year-old engineer was quick and eager. As the men talked, the future airplane began to take shape on paper. To lift the great fuel load, the wings would have to be lengthened. But by how much? Neither of them even knew exactly how far it was to Paris from New York and thus how much fuel would be needed. A trip to the library answered the first question. By stretching a string around a globe, they reckoned the distance to be 3,600 miles. Lindbergh proposed a range of 4,000 miles for the new plane. Given the Wright Whirlwind's rate of fuel consumption, he would need at least 400 gallons; 425 would be safer. A 46-foot wing could lift the resulting weight. A huge fuel tank would have to be installed under the wing; it must be near the center of gravity, so the plane would remain balanced as fuel was consumed.

Hall was surprised to learn that Lindbergh intended to fly alone, but he immediately saw the advantage it offered. Lindbergh, for his part, was becoming a fanatic about weight. Every ounce trimmed from the plane made the fuel load less burdensome. Dispensing with a navigator would save about 350 pounds.

Lindbergh wanted the cockpit built behind the fuel tank so that he would not be crushed between it and the engine if he crashed. But sitting behind the tank meant he could not see forward. Lindbergh considered this a small loss. Over the Atlantic there would be nothing to run into; on takeoff and landing, he could turn the plane slightly sideways to keep an eye on the runway. Most pilots of the day sideslipped into small fields anyway, giving themselves a clear view until the last moment before touchdown. Eventually a small periscope would be installed so that Lindbergh could look ahead, but he rarely used it.

"What are you going to use the plane for later on?" Hall asked Lindbergh, knowing that the blind cockpit arrangement would lessen its commercial value.

"I haven't even thought about what to do with the plane after I land at Paris," Lindbergh said, "and I'm not going to until I get there."

Hall wanted to enlarge the M-2's tail surfaces to improve stability; the cost would be a slight increase in drag, and thus a small loss of range.

"Let's put everything into range," Lindbergh countered. "I don't need a very stable plane. I don't plan on going to sleep."

The plane would be precisely tailored to the man who would fly it. Hall even designed the cockpit to Lindbergh's dimensions and calculated the pilot's weight as Lindbergh's: 170 pounds. The price, including a Wright Whirlwind J-5C engine, came to $10,580. On February 25, Lindbergh made a down payment of $1,000 and Ryan Airlines started work on the *Spirit of St. Louis.*

As construction shifted into high gear in San Diego, news stories about Lindbergh's competitors on the East Coast began to appear. Richard Byrd announced his plans for the big Trimotor that Fokker was building in New Jersey. Noel Davis and his copilot, Stanton Hall Wooster, had completed preliminary testing of their three-engined *American Legion.* Charles Levine had ordered large fuel tanks installed in the Bellanca *Columbia,* though he kept his plans a crazy quilt of contradiction and rumor to squeeze publicity out of the press. Lindbergh officially entered the Orteig competition on February 25 and saw his name misspelled "Linberg" or "Lindenburg" in newspaper reports that discounted his chances. In France, Nungesser and his navigator, François Coli, were ready to depart for New York in *L'Oiseau Blanc—"The White Bird"—* at the first sign of a break in the prevailing westerly winds.

As spring bloomed, Noel Davis appeared to have the best chance to win the prize. With a series of faultless flights already behind him, he flew the silver and bright yellow *American Legion* to Langley Field, a military installation in Virginia, for the final tests, gradually working up to a takeoff with a full load of fuel. At last he was ready. Early on April 26, Davis and Wooster confidently fired the plane's three Whirlwind engines. Topped off with fuel, the *American Legion* weighed 17,000 pounds, but Davis was unperturbed. The runway was nearly a mile long and unobstructed except for a row of trees well beyond its end. Just the night before he had declared, "I'm sure we've licked all our problems."

Davis waved the chocks away and Wooster advanced the throttles. The three Whirlwinds thundered for a moment at the beginning of the runway. Then the plane moved slowly away from a group of watching officers, some of whom had begged to be taken along. It lumbered along the smooth ground, its tail lifting as it picked up speed. When it was well down the field, far beyond the point at which Davis had expected to be airborne, the plane at last lurched into the air.

Wooster held the plane low to gain speed, then tried to pull up over the trees that now loomed close ahead. The heavily laden craft rose almost imperceptibly. The spectators sensed that the plane was in trouble. They saw it turn slightly to the right, evading the trees. But the turn slowed its speed and it began to lose altitude. Ahead lay a marsh and beyond that a broad pool. A man fishing there saw the plane, heard the engines suddenly go silent and thought the craft was going to hit him. Instead, the wheels plowed a furrow in the marsh, throwing a spray of water. The plane's momentum carried it into the pool, where it slammed into the bank on the far side, tail erect, cockpit buried in mud. When rescuers reached the scene, they found Wooster and Davis dead. The *American Legion* had risen no more than 20 feet off the ground.

Crashes plagued the Orteig Prize contenders. Ten days before Davis' fatal mishap, Anthony Fokker himself took the Trimotor his company had built for Commander Byrd on its maiden flight from the airport in Teterboro, New Jersey. Fokker had planned to fly alone, but at the last moment he took on some unexpected passengers: Byrd, his pilot, Floyd Bennett, and a radioman. Their combined weight made the plane nose-heavy, but the flight proceeded uneventfully until Fokker came in to land. After the wheels touched the runway, the plane sped along for 200 feet or more with only the two front wheels on the ground. "The tail," according to one observer, "kept rising and rising until finally she turned over in a complete somersault." Fokker was thrown free, shaken but uninjured. Bennett suffered a broken leg, Byrd's arm was fractured and the radioman sustained internal injuries. Oddly, the plane was not seriously damaged; it was repaired within three weeks.

After crashing on takeoff for its final test flight, the American Legion rests nose down in a marsh near Langley Field, Virginia. Both the plane's pilot and copilot, Lieutenant Commander Noel Davis and Lieutenant Stanton Wooster, were killed.

In the Wright-Bellanca camp, Chamberlin was almost ready to go. With Bert Acosta, a veteran barnstormer and racer, as copilot, he had taken the plane up to test its endurance. They stayed aloft for 51 hours 11 minutes, a new world record—and more than long enough to reach Paris. Shortly thereafter, Chamberlin gave owner Charles Levine's daughter and another child a ride in the *Columbia.* As the plane took off, the left wheel came lose. Apprised of the problem by a pilot in a chase plane, who got the message across by waving a tire, Chamberlin made a superb emergency landing, touching down on the right wheel and wing tip. The plane ground-looped, but no one was injured and the plane was airworthy again in a matter of days.

In San Diego, meanwhile, the major components of the *Spirit of St. Louis* began to take shape, with welders assembling the fuselage on the first floor of the cannery building and carpenters building the wing in the loft overhead. Lindbergh prowled about, wearing a blue business suit, insisting that the workmen tailor the plane to his most minute specifications. He worried, for example, about the lines that ran between the throttle and the engine. They were made of stiff piano wire that slid inside guides of lubricated ¼-inch steel tubing. The guides were painstakingly shaped by hand, each bend in the control-wire routes carefully crafted and anchored to the plane. Lindbergh thought that the tubing was too narrow and that the control wires might bind. So the workmen redid the job—with tubing 1/16 inch wider. Watching a length of copper tubing being shaped into an oil line, Lindbergh ordered: "Break the oil lines every 18 inches and reconnect them with rubber hose."

"What for?" demanded Ed Morrow, a forthright Oregonian who worked for Ryan as a metal worker and draftsman.

"Well," replied Lindbergh, "most forced landings I've heard about on long hops were caused by a break in the oil line. From vibration. If we put rubber hose connections in, they'll absorb the vibration." The change was made.

When the mechanics were ready to mount the engine—one of four specially inspected Whirlwinds that had been supplied on consignment by the Wright Corporation—Lindbergh was there. "We gather around the wooden crate," he wrote later, in the present tense he employed to transport himself back in time, "as though some statue were to be unveiled. It's like a huge jewel. We marvel at the quality of Cosmoline-painted parts. On this intricate perfection I'm to trust my life across the Atlantic Ocean."

Once the engine was in place, someone accidentally dropped a wrench on it, chipping a cooling fin. "Lindbergh almost cried," O. L. Gray, an installation mechanic, recalled. "I told him we could smooth it out with a file and a little paint and never know the difference."

"I'll always know the difference," Lindbergh replied. "I want another engine in there."

"Why does this damn thing have to be so perfect?" another of the workmen muttered.

"For two reasons," rejoined Lindbergh. "The first is, I'm not a good swimmer. The other reason is that I've always been taught that perfection pays off." The engine was replaced.

Lindbergh was as demanding of himself as he was of the Ryan crew. Two matters concerned him: navigation and saving weight. Reluctant to ask for help in laying out a course to Paris—he preferred not to expose himself as an inexperienced navigator—he plotted out a great circle route, the shortest distance between two points on the globe; it would require him to alter his heading slightly every hour during the flight to maintain his course. To keep the airplane as light as possible, Lindbergh decided to fly without a radio, saving 90 pounds. He would take no sextant, because he could not use one and fly at the same time. Nor would the *Spirit of St. Louis* have fuel gauges, which were heavy and usually inaccurate anyway. He would keep track of fuel consumption with a watch and a tachometer, the instrument that showed engine speed. He went so far as to refuse $1,000 from a stamp collector to carry one pound of letters to Paris.

Given Lindbergh's carefulness, it is just as well that he was away from the Ryan plant the night that some exhausted workers, who often toiled past midnight, uncorked a bottle of wine. Presently Hawley Bowlus, the factory manager, picked up one of the girls who had been sewing fabric for the plane.

"You've been working too hard, Peggy," he announced jovially, and set her on the wing, which still lay on the assembly table. The staccato sound of wing ribs cracking sobered the group instantly. They repaired the damage by morning, and no one told Lindbergh about the mishap until years later.

After nearly two months, during which no one at Ryan Airlines worked on anything but Lindbergh's plane, the wings and fuselage were ready to be taken to Dutch Flats Airfield, four miles away, for assembly. Almost at once the transfer hit snags. To support the 2,500-pound load of fuel the plane would start out with, Hall had widened and strengthened the undercarriage, basing his design on one "borrowed," with the aid of a flashlight in the dead of night, from a Fokker F.VII hangared nearby. With the new landing gear attached, however, the fuselage would not fit through the factory door, so one side of the undercarriage had to be removed. In addition, the wing was too long to be carried downstairs from the loft. Instead it had to be passed through a window and onto the top of a railroad boxcar pushed alongside the building, and thence lowered to the ground. Lindbergh, as usual, was there to help.

Once these maneuvers were accomplished, the components were trucked to the airfield and put together. On April 28, sixty days after work began, the plane was rigged and ready to fly. "What a beautiful machine it is," Lindbergh marveled, "trim and slender, gleaming in its silver coat."

Lindbergh made the test flight, of course. "We thought he would lift off, wobble the controls, see how she felt and come in," said Ed Morrow, the draftsman. "But no—he took off and went right on up." Lindbergh put the plane through its paces, logging 128 miles per hour in a speed trial. He had a hot airplane on his hands that demanded constant attention. As Don Hall had predicted, the tail surfaces were too small in relation to the wingspan, and the plane was somewhat unstable. But Lindbergh had expected that. When a curious Navy pilot approached in a Curtiss Hawk fighter, Lindbergh, the ex-Army aviator, engaged the intruder in mock combat. Up and around they zoomed, each trying to line up the other in imaginary gun sights. The result was a draw; the Hawk had more speed but the new Ryan was more maneuverable.

Test flights continued on subsequent days with the tanks increasingly full. But at 300 gallons, less than three quarters of the load the plane would have to carry on the flight to Paris, the landing gear—even though it had been reinforced—was visibly strained. Lindbergh, fearing that more weight might break the undercarriage on landing, decided to test no more.

Then suddenly it appeared that the whole effort might have been made in vain; word flashed from Paris on May 8, 1927, that Charles Nungesser and François Coli had taken off for New York.

The 46-foot wing of the Spirit of St. Louis dangles from a derrick while Lindbergh (arrow, center) helps push a railroad car out of the way. The top of the boxcar served as the first step down from the second story loft (rear) where the wing was built.

Rivals in the great race

From the summer of 1926 until the summer of 1927, six airplanes became involved in the intense race to be the first to fly nonstop between New York and Paris. The winner, Charles Lindbergh's *Spirit of St. Louis,* is shown on pages 74-75; the also-rans, each distinguished in its own right, are presented here in scale with one another.

The six contenders exemplified competing aeronautical theories: Three were monoplanes, three biplanes; three had single engines, three multiple engines; five had air-cooled engines, the other water-cooled. Only three made it across the ocean: Richard Byrd's *America,* Clarence Chamberlin's *Columbia,* and of course, the *Spirit of St. Louis.* René Fonck's *New York/Paris* and Noel Davis' *American Legion* crashed on takeoff at a cost of four lives. *L'Oiseau Blanc,* the French entry of Charles Nungesser and François Coli, disappeared over the North Atlantic.

SIKORSKY S-35—*NEW YORK/PARIS* (FONCK)

Largest of the contenders, the S-35 had a 101-foot wingspan and weighed more than 14 tons when its wing tanks and the two 1,100-gallon tanks behind its outboard engines were full. It was powered by three 450-hp Jupiter engines.

KEYSTONE PATHFINDER—*AMERICAN LEGION* (DAVIS)

The ill-fated American Legion, its fuselage boldly outlined by a black stripe, was a converted bomber with three Wright Whirlwind engines.

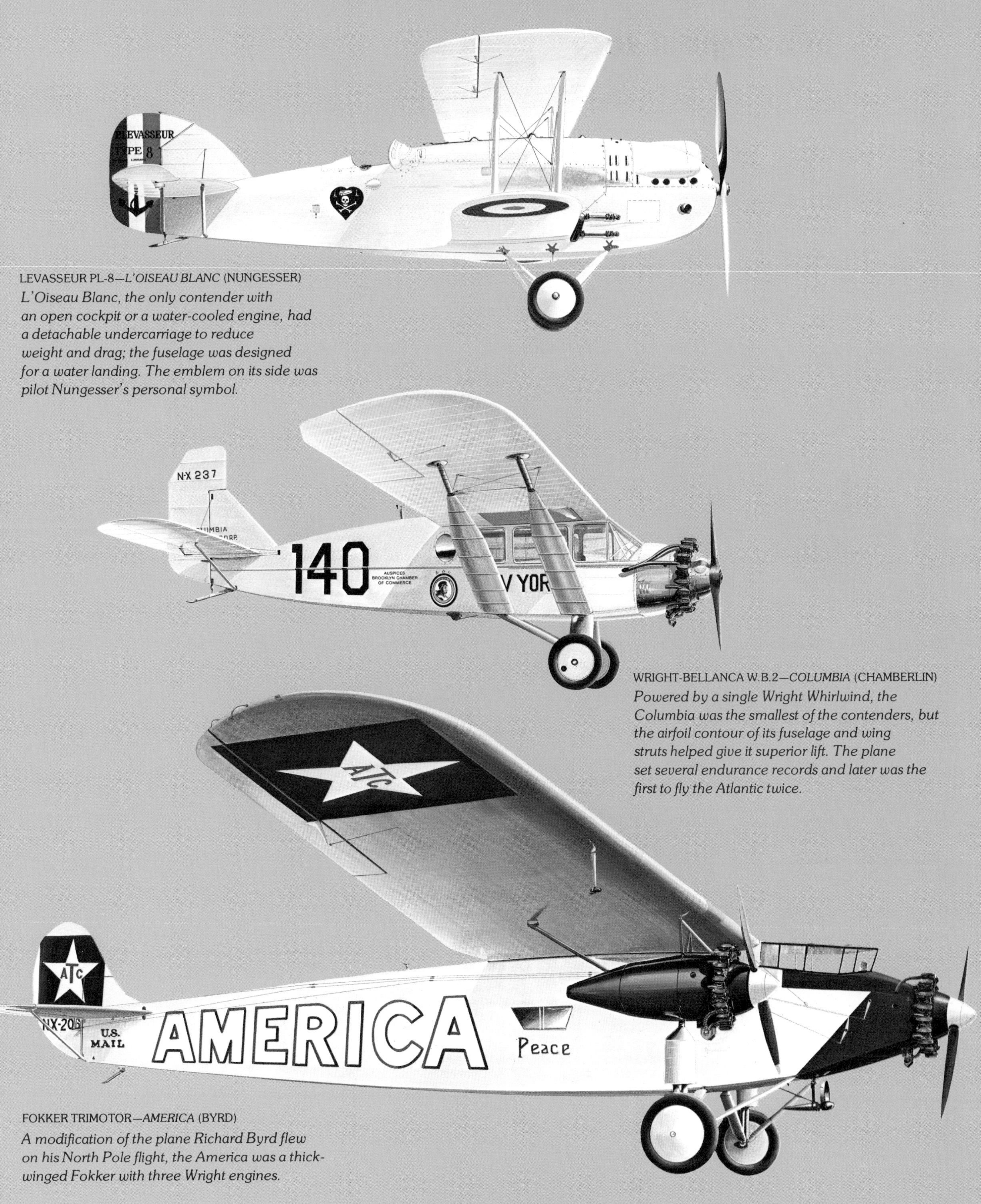

LEVASSEUR PL-8—*L'OISEAU BLANC* (NUNGESSER)

L'Oiseau Blanc, the only contender with an open cockpit or a water-cooled engine, had a detachable undercarriage to reduce weight and drag; the fuselage was designed for a water landing. The emblem on its side was pilot Nungesser's personal symbol.

WRIGHT-BELLANCA W.B.2—*COLUMBIA* (CHAMBERLIN)

Powered by a single Wright Whirlwind, the Columbia was the smallest of the contenders, but the airfoil contour of its fuselage and wing struts helped give it superior lift. The plane set several endurance records and later was the first to fly the Atlantic twice.

FOKKER TRIMOTOR—*AMERICA* (BYRD)

A modification of the plane Richard Byrd flew on his North Pole flight, the America was a thick-winged Fokker with three Wright engines.

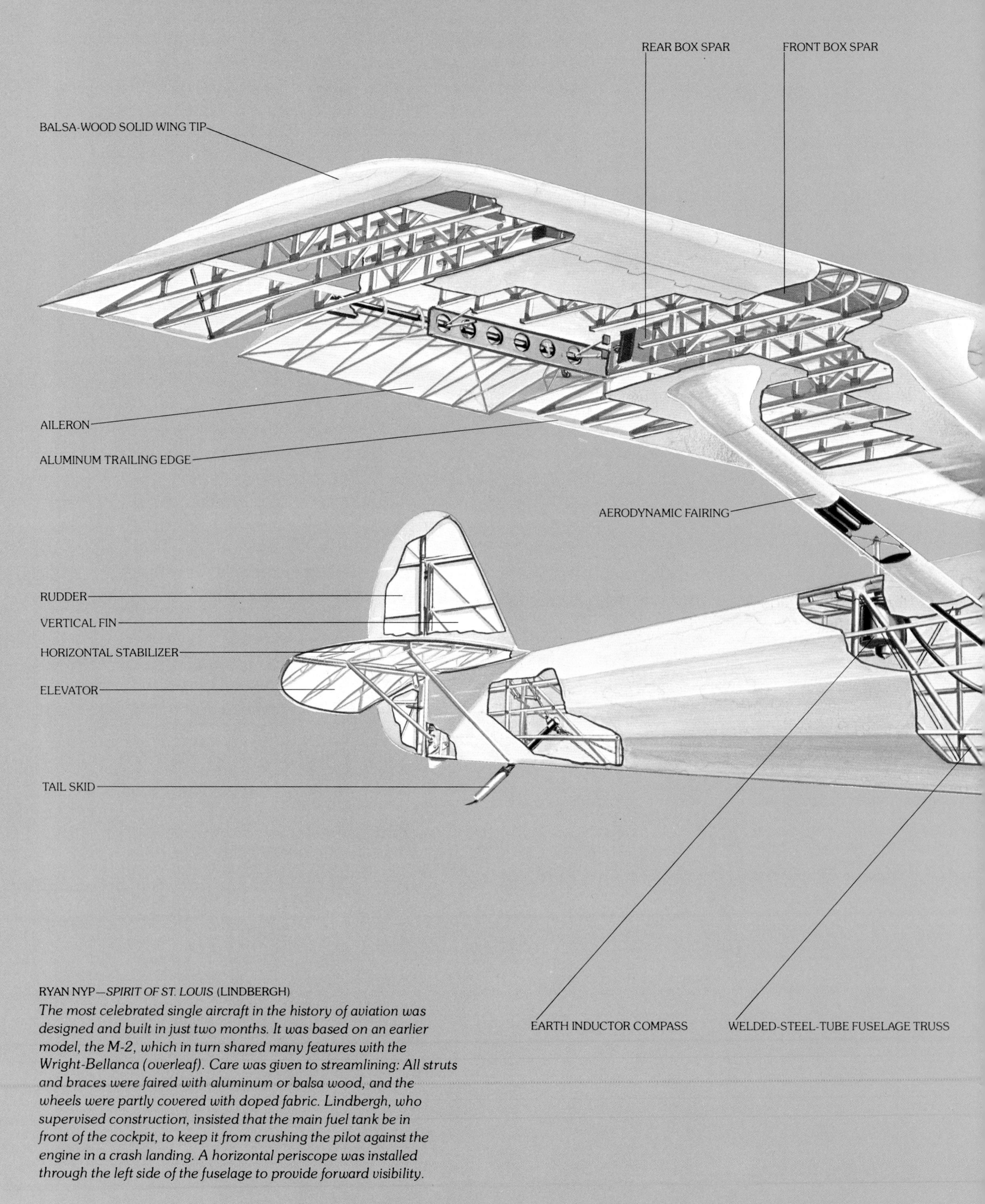

RYAN NYP—*SPIRIT OF ST. LOUIS* (LINDBERGH)

The most celebrated single aircraft in the history of aviation was designed and built in just two months. It was based on an earlier model, the M-2, which in turn shared many features with the Wright-Bellanca (overleaf). Care was given to streamlining: All struts and braces were faired with aluminum or balsa wood, and the wheels were partly covered with doped fabric. Lindbergh, who supervised construction, insisted that the main fuel tank be in front of the cockpit, to keep it from crushing the pilot against the engine in a crash landing. A horizontal periscope was installed through the left side of the fuselage to provide forward visibility.

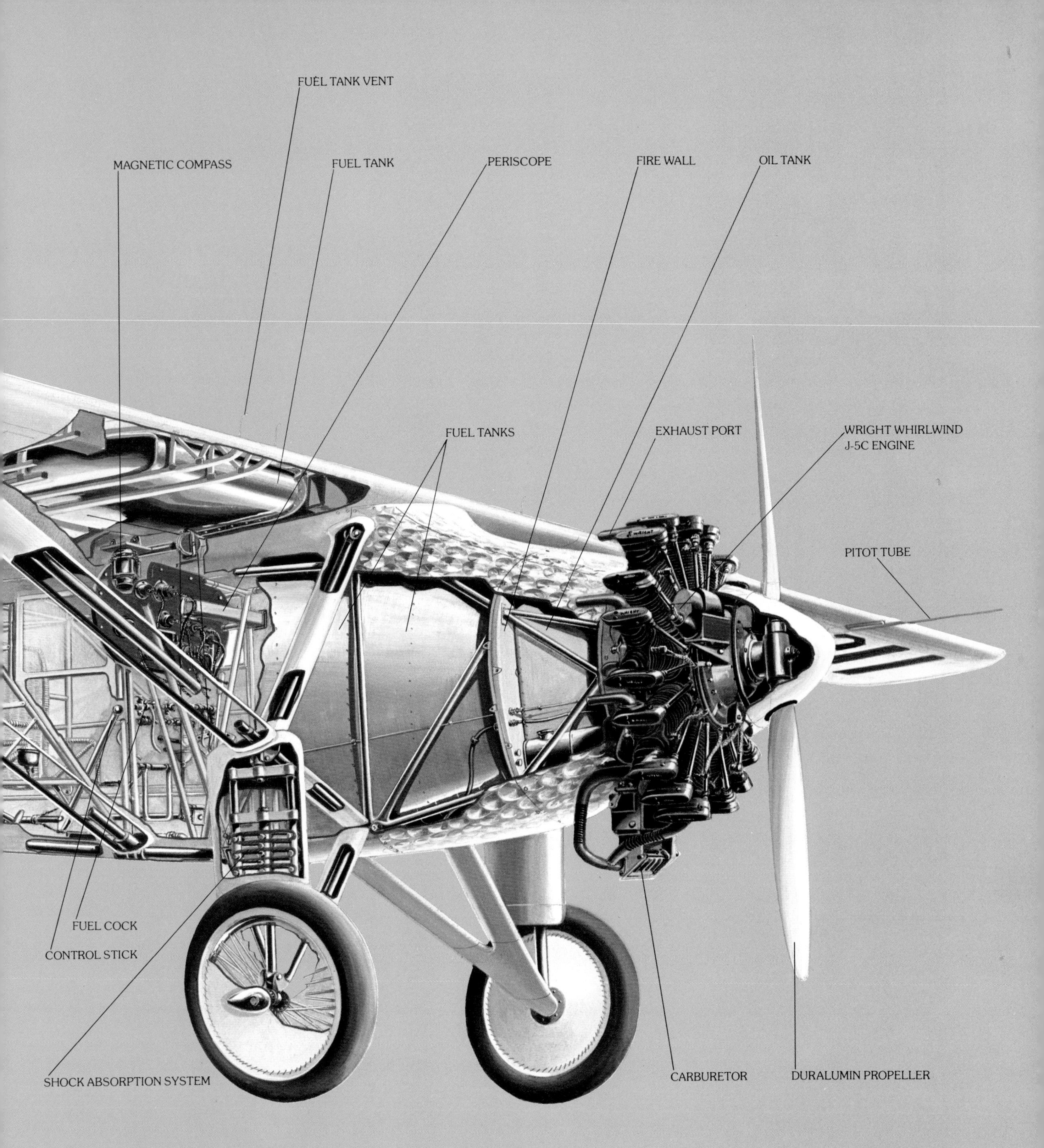
FUEL TANK VENT
MAGNETIC COMPASS
FUEL TANK
PERISCOPE
FIRE WALL
OIL TANK
FUEL TANKS
EXHAUST PORT
WRIGHT WHIRLWIND J-5C ENGINE
PITOT TUBE
FUEL COCK
CONTROL STICK
SHOCK ABSORPTION SYSTEM
CARBURETOR
DURALUMIN PROPELLER

N-X-211

3

Tragedy and triumph

When Charles Nungesser took off from Paris in his bid to cross the Atlantic, he was making a last-chance attempt to fly his way back into the hearts of his countrymen. Paris-born, Nungesser grew up exuding heroic potential. He was a handsome, muscular youth who delighted in boxing, swimming and racing motorcycles. At 16 he quit school and his parents staked him passage to Brazil to visit an uncle who had emigrated there.

The uncle proved elusive; he had moved to Argentina and it was five years before Nungesser found him, living on a sugar plantation. In the meantime, young Charles had worked as a mechanic and had become well known as a race-car driver. He had achieved further celebrity in Buenos Aires by knocking out a local prize fighter in a challenge match after weathering a severe beating in the early rounds.

He also taught himself to fly, accomplishing the feat in one daring lesson. Visiting an airfield in Argentina one day, Nungesser asked the pilot of a plane that had just landed to let him fly it. The pilot's amused refusal angered Nungesser, and convinced from conversations with other pilots that any good automobile driver could fly a plane, he scrambled into the cockpit and took off. To the amazement of those watching, he flew around for several minutes getting the feel of the controls, then landed safely, if bumpily. The experience confirmed his desire to fly and after two weeks of intensive lessons—with an instructor—he went barnstorming with the same pilot whose plane he had borrowed.

Drawn back to France early in 1914 by premonitions of war, Nungesser signed on with the cavalry, and when the War began he quickly distinguished himself as a fearless, if somewhat madcap, fighting man. Trapped behind German lines with two infantry stragglers, he ambushed a German command car, killed everyone inside, then drove through a hail of bullets from both sides—Germans firing at French uniforms, French firing at the German car—to rejoin his regiment.

Shortly after this episode, he managed a transfer into the air service and hurled himself into aerial combat with such vigor that he seemed at times to lack the normal human instinct for self-preservation. Regularly he brought his plane home perforated by bullets; once, after he crash-landed behind French lines, mechanics patched more than 40 holes in his Nieuport. During three years of combat flying, he shot down 45 German planes, but he was wounded himself no fewer than 17 times. Surgeons had to replace some of his bones with plates and chunks of

The Spirit of St. Louis, photographed from a press plane just minutes after takeoff from Roosevelt Field on Long Island, soars over a tranquil forest-fringed lake, carrying its young pilot, Charles Lindbergh, to Paris and glory.

precious metals, including a silver ankle—to which he credited much of his postwar skill at dancing the Charleston. Awarded the French Légion d'Honneur and a dozen other medals from eight grateful Allied nations, Nungesser emerged from the War a hero, loved perhaps not so much for his medals and victories as for his indomitable spirit, which drove him into battle even though he sometimes had to be lifted to his plane.

For a time after the Armistice, he combined fame with a bon vivant's charm to live in the style he thought appropriate for a national hero. He took rooms at an expensive Paris hotel and drove a Rolls-Royce given to him by the British government in recognition of his combat record.

But France wanted to forget the War; gradually Nungesser became less fashionable. Fewer hostesses regarded him as an intriguing dinner guest. His frequent amorous conquests, once overlooked or quietly applauded as the way of a hero, now earned him an awkward notoriety. By the early 1920s, a tide of debt threatened to engulf him. He opened a flying school at Orly, outside Paris, but it failed, as did his marriage to an American woman, Consuelo Hatmaker. Nungesser next toured the United States with a French flying circus that re-created his great air battles in some 50 cities. When the circus disbanded he starred in a silent film, *The Sky Raider,* and toured the country to promote it.

In 1926 he headed back to France, telling friends "The next time I return here will be by air." He did not say it lightly; to fly the Atlantic was an enduring dream of Nungesser's. To accomplish it would transform him from a has-been to a hero again. René Fonck's crash on takeoff from New York was still reverberating in the French press when Nungesser walked into the Paris office of Pierre Levasseur, a well-known aircraft designer. Nungesser presented Levasseur a detailed plan for the same flight, but in the opposite direction. Nungesser contended that it was ridiculous for a Frenchman to sail to America in order to fly to France. Besides, starting the flight in Paris would save the cost of shipping the plane. It was an economy that appealed to a consortium of Parisian sportsmen who still admired Nungesser and had agreed to back him. Flying west instead of east, however, a pilot must expect head winds most of the way; that would require additional fuel and extend the flight time at least to 40 hours. Like Lindbergh, Nungesser knew the merits of a single-engined plane that he could fly alone, and Levasseur's PL-8, designed for the French Navy, seemed best suited to the task.

Impressed with Nungesser's plan, Levasseur agreed to tailor a PL-8 to his requirements. He had, however, one firm suggestion: Nungesser should not make the attempt alone but should take a navigator. Gradually Nungesser came around to the designer's point of view. The head winds would consume so much fuel that there would be precious little in reserve to compensate for errant navigation.

As his navigator, Nungesser chose Captain François Coli, 45, who wore a black patch to cover the loss of an eye in the War. Coli was an obvious choice. Born to a seafaring family in Marseilles, he was a master maritime navigator who had switched to the air service and had become

Charles Nungesser's World War I uniform is laden with medals from eight nations in this photograph, which was taken to publicize The Sky Raider, a movie he made in Hollywood in 1923, four years before he attempted the Atlantic crossing.

commander of a fighter squadron. After the War, he piloted the first round-trip flight across the Mediterranean and the first nonstop flight from Paris to Casablanca. Coli shared with Nungesser a lasting urge to fly the Atlantic. He and another partner had registered for the Orteig Prize in 1925, but their plans had been disrupted by a crash that demolished their plane.

Nungesser's new PL-8 took shape during the winter and by spring it was unveiled. Named *L'Oiseau Blanc,* it was a handsome white biplane with a watertight fuselage made of sealed plywood. After takeoff it would jettison its wheels to increase speed and range, then land on its boatlike belly in New York Harbor at the end of the trip. The plane weighed about twice as much as the one Ryan Airlines was building for Lindbergh and it had an engine twice as powerful—a 12-cylinder water-cooled 450-horsepower Lorraine-Dietrich that could maintain a cruising speed of 100 miles per hour. It also carried about twice the fuel of Lindbergh's plane, 886 gallons, in three aluminum-alloy tanks centered between the wings. But unlike Lindbergh, Nungesser would be able to see where he was headed; an open double cockpit was positioned just aft of the wings.

In spite of the best efforts of Nungesser and Coli to conceal their plans and prepare for the flight undisturbed by public clamor, word leaked out, and the press began to hound them. But the usually vainglorious Nungesser now had no interest in publicity or in accommodating those friends who had so recently abandoned him. He and Coli dodged reporters and well-wishers to focus instead on meteorology, sea charts, wind currents, routes, fuel consumption rates, load balance and emergency equipment. They pondered whether to take a radio, and decided to do without one to save weight. They exercised with bar bells and medicine balls and practiced staying awake for long periods. Coli bragged that he could go without sleep for 60 hours at a time.

In early April, Nungesser flight-tested the plane at Villacoublay, on the outskirts of Paris. *L'Oiseau Blanc* maneuvered well, climbed readily to 18,000 feet and clocked 128 miles per hour in a speed trial. Nungesser decided not to risk a takeoff with full tanks until he was ready to go. "Either we will get off and the test won't mean anything," he explained, "or we won't get off—and there'll be plenty of opportunity to learn about that when we go."

The tests drew increasing public attention, the only gratifying aspect of which was a flood of offers to purchase their stories should the flight succeed. Coli had shared Nungesser's postwar financial difficulties. "You see," the pilot told his navigator, "we will soon have more money than we need to pay our debts and live to be old." They had not formally registered for the Orteig Prize, but undoubtedly expected that technicality to be waived after they made the flight.

The men and the plane were ready, but unfavorable winds and weather delayed their departure for weeks. Nungesser and Coli visited the meteorological offices on Avenue Rapp every day to study the disappointing charts. Coli had plotted three alternative routes, and the weather remained miserable on all of them.

Then in rapid succession came word of Byrd's and Davis' crashes. The news sparked doubt that Nungesser could succeed. But when the papers reported that Chamberlin was ready to leave New York at any moment in his Bellanca, renewed public pressure fell on Nungesser to beat the Americans into the air. The old hero barked at reporters: "I am aware that each night in well-known Paris bars numerous ground aviators successfully cross the Atlantic between cocktails, but I have been carefully preparing for three years." He was not about to be pushed.

On May 7, Nungesser and Coli moved their plane to Le Bourget airfield, where the two-mile dirt runways had been specially tamped to iron out bumps. An excited crowd that had gathered at the airfield surged through police lines toward the aircraft, but order was restored when 50 soldiers, their bayonets fixed, reinforced the police. On the day of the transfer a rumor spread that Chamberlin had taken off. Dismay permeated Nungesser's camp until the Chamberlin story proved to be a false alarm. Later that day, as a storm gathered over Paris, the chief meteorologist reported that a rare east wind was developing: "The situation is extremely favorable," he said. "The storm in Paris is purely local and for 2,000 kilometers you will fly with the wind behind you. Near Newfoundland conditions are a little less propitious, but it seems to me that you will wait a long time before you have as good conditions."

"We're off," announced Coli.

"Agreed," said Nungesser, and the two men shook hands. Mechanics began fueling *L'Oiseau Blanc.*

An old squadron mate turned up to wish them well. "How do you count my chances?" Nungesser asked. "Because it is you," answered the pilot, "I grant you a 10 per cent chance of success. If you do not end up in the juice, it is because you are named Nungesser!"

The storm seemed to hold off as the crowd, many of them in evening clothes, grew in size. They carried food hampers, and corks were soon popping from bottles of champagne. Restaurants prepared to stay open all night. Many notables joined the throng; President Gaston Doumergue arrived, and the boxer Georges Carpentier, whose fight with Jack Dempsey had attracted the first million-dollar gate, came with an entourage. Nungesser and Coli ate at a table well away from the crowd, then slept for two hours. They awoke at 3 o'clock on May 8, 1927.

When Levasseur, the designer, arrived at the airfield, Nungesser asked, "No news of Chamberlin?"

"Still nothing," came the reply.

As the sky paled with dawn, they drove to the plane in an open car, and the crowd burst into cheers. *"Allons!"* shouted Nungesser. The single engine roared and the crowd saw his lips form the words *"Au revoir."* It was 5:17 a.m. The crowd fell silent as the plane rolled down the long runway, gathering speed. Once it tried to take off, rose slightly and sagged back to earth. There were cries of alarm. Then, nearly a mile

Just before embarking on their ill-fated flight from Paris to New York, Charles Nungesser and his navigator, François Coli—who had lost an eye during World War I—strike a confident pose in the cockpit of their plane, L'Oiseau Blanc. The macabre insignia on the fuselage also distinguished Nungesser's aircraft during World War I, when he became famous for taunting death as a fighter pilot.

down the runway, the plane lifted off and began to climb. Several minutes later, Nungesser dropped the landing gear to gain speed and altitude. At 6:48 a.m., *L'Oiseau Blanc* crossed the French coastline and disappeared into the clouds over the Channel.

Five hours later Nungesser and Coli were sighted off Ireland, but contrary to the forecast, they were flying into the wind. Meteorologist James Kimball, weather adviser to the Orteig competitors in New York, estimated that Nungesser was running into head winds of 25 miles per hour over the Atlantic. Winds of that velocity would reduce *L'Oiseau Blanc's* cruising speed to little more than 75 miles per hour. If the wind persisted, the plane, though carrying fuel for 43 hours of flight, would run out 400 miles short of the mark.

France slept fitfully on the night of May 8 and awoke eager for news of the two fliers. About 10 a.m. Paris time, word came that they had been sighted over Newfoundland. The United States Navy reportedly tracked Nungesser and Coli over Portland, Maine, at 2:53 p.m., then over Boston. It seemed that Nungesser and Coli would become heroes in New York about the time Parisians started home from work.

A little after 5 p.m., Paris was galvanized by news that *L'Oiseau Blanc* had landed safely in New York harbor. The newspaper *La Presse,* in an extra edition, bannered a complete description of how an aircraft escort met the Frenchmen as they approached the city, how the fliers embraced in the cockpit, how an immense crowd sprinkled with notables had assembled to congratulate them. At least one other paper, without checking, repeated the story. Paris went wild. France had triumphed!

In New York, however, the scene was far different from the one described in the French newspaper. A crowd had gathered, to be sure, but it saw no embrace, no escort of planes. There was in fact no sign of Nungesser and Coli. *La Presse* had simply made up the story in order to be out with it first. At dusk the crowds melted away from New York's Battery Park. The hour had passed when *L'Oiseau Blanc's* fuel would be exhausted; the plane was down somewhere. A giant search began.

In San Diego, the news of Nungesser's departure hit Lindbergh and the crew at Ryan Airlines like a body blow. Lindbergh remained composed, but when Don Hall, the designer, blurted out, "I almost hope they don't make it," Lindbergh flushed angrily. "Don't say that," he snapped.

For almost a week the *Spirit of St. Louis* had been ready to fly to New York. A low pressure area, which blanketed the Rocky Mountains to the east in clouds and rain, had kept Lindbergh on the ground. On May 9, the day that Nungesser and Coli disappeared, Lindbergh received the good news that a high pressure area was moving in from the west; it would clear the mountains and drive the storm eastward. He decided to follow the high. If he stayed close behind it, he reasoned optimistically, he might reach New York just as it passed, then follow it across the Atlantic before Chamberlin or Byrd could take off.

The next morning he stopped by the Ryan plant to say goodbye. The

25 Centimes TROISIÈME ÉDITION 25 Centimes

L'INTRANSIGEANT

Directeur : LÉON BAILBY

Mardi 10 Mai 1927

TROISIÈME ÉDITION

LE GEL ET LES PLANTES

Sommes-nous en lune rousse?

Commencez à la deuxième page, un feuilleton, les souvenirs de Maurice CHEVALIER

FLEURS DE MAI

A trois ans, elle arrivait seule d'Amérique

TROISIÈME ÉDITION

NUNGESSER est arrivé

Après avoir annoncé le passage de l'avion au-dessus de Terre-Neuve et au large d'Halifax (Canada), on signale son arrivée en vue de New-York

CEUX QUI PORTENT NOS COULEURS

DERNIÈRE MINUTE

DEUX NOUVELLES DÉPOSITIONS

L'espionnage communiste

A quoi peut servir la valise diplomatique de l'ambassade des Soviets

ANNIVERSAIRE

DERNIÈRES NOUVELLES

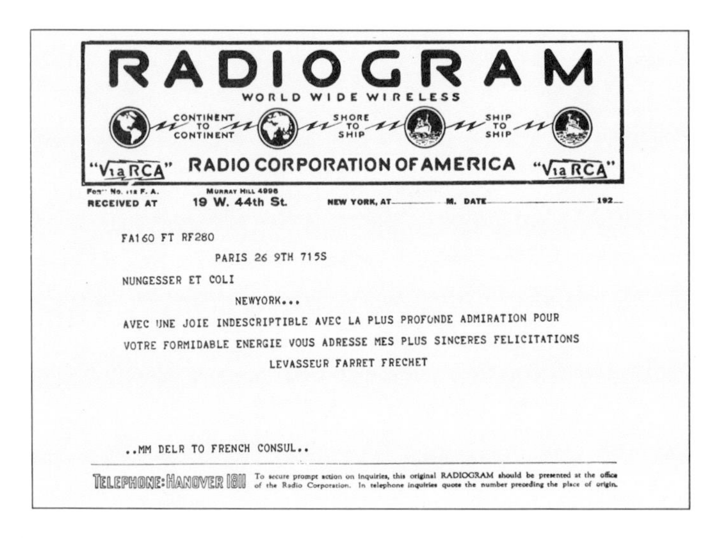

RADIOGRAM

WORLD WIDE WIRELESS

CONTINENT TO CONTINENT · SHORE TO SHIP · SHIP TO SHIP

"Via RCA" RADIO CORPORATION OF AMERICA "Via RCA"

RECEIVED AT 19 W. 44th St. NEW YORK, AT ___ M. DATE ___ 192__

FA160 FT RF280

PARIS 26 9TH 715S

NUNGESSER ET COLI

NEWYORK...

AVEC UNE JOIE INDESCRIPTIBLE AVEC LA PLUS PROFONDE ADMIRATION POUR VOTRE FORMIDABLE ENERGIE VOUS ADRESSE MES PLUS SINCERES FELICITATIONS

LEVASSEUR FARRET FRECHET

..MM DELR TO FRENCH CONSUL..

TELEPHONE: HANOVER 1811 To secure prompt action on inquiries, this original RADIOGRAM should be presented at the office of the Radio Corporation. In telephone inquiries quote the number preceding the place of origin.

False reports of a successful transatlantic flight by Frenchman Charles Nungesser sparked a premature Paris celebration; actually, he and his navigator, François Coli, were lost at sea. A bogus headline (top) trumpets Nungesser's arrival in New York, while a message (above) from the plane's designer, Pierre Levasseur, expresses "indescribable joy and profound admiration." At left, Madame Laure Nungesser sits in her home surrounded by mementos of her dead son.

workers who had built the plane had signed their names on a wing spar, before the fabric covering was added. Ed Morrow remembered that Lindbergh remarked lightly, "I might get wet, you know."

"Not after we've worked so hard on that plane," Morrow responded. "If it don't get wet, you won't."

Lindbergh smiled; he knew the risks. A few days earlier Claude Ryan had asked him what he thought his chances were.

Lindbergh turned the question back, rephrasing it. "What do you think the percentage is?"

Ryan, trying to sound optimistic, replied, "I think you've got a 75 to 25 per cent chance."

Lindbergh grinned. "That's just about what I figured."

At 3:55 p.m. on May 10 he was off for the East Coast. The plane carried little more than half a load of fuel, enough to reach St. Louis, the only stop he planned on the way to New York. Lindbergh cruised eastward across the desert, following the Atchison, Topeka and Santa Fe railroad toward the mountains and the darkness. At sunset, 4,900 feet high and climbing, he corrected his heading by the golden gleam of the Sante Fe tracks below. Later, at 8,000 feet, Lindbergh felt the cold of the night and the altitude. A moon shone, but haze made the mountains of Arizona below indistinct. There were no lights to guide him.

Then his engine began to miss. It sputtered again and again. Engine trouble, at night, in poor visibility, over mountains. It was an alarming combination of bad luck. He pulled back the throttle, adjusted the fuel mixture, checked the fuel pump, tested both magnetos. Nothing seemed wrong, but the uneven pulse continued, shaking the whole plane. He was losing altitude. It occurred to him that he was finished before he had really started. Forget the dream. Lacking a parachute, he would be lucky to save his skin; and the plane would probably be wrecked on the slopes below. He circled, studying the ground, and saw an untimbered meadow, visible as a light patch between wooded areas that reflected less moonlight. He did not know what boulders and gullies it contained, but at least it was open ground. He would go in there.

Then Lindbergh realized that he was sinking more slowly than he had thought. The engine still delivered some power. He continued to adjust throttle and mixture, trying to reduce the shuddering that racked the plane. He stopped losing altitude and continued to circle. Maybe the engine would not quit. Perhaps the problem was carburetor ice that would melt as he dropped lower. He would have a carburetor heater installed in New York. If he ever got to New York.

The engine began to run more smoothly and he found that he could climb again. His confidence began to return. But ahead lay mountains more than 12,000 feet high. Lindbergh faced a choice. He could turn back, but miles of rugged terrain stretched between him and the Coast, which by now might be fogged in. He could continue to circle until dawn, tethered to that ambiguous clearing below—slight security but better than nothing if the engine died. Or he could press on.

Nothing defines Lindbergh or explains the outcome of his mission more clearly than his decision that night to continue alone over inhospitable mountains in a tiny plane with a sputtering engine. He resumed his original course and began to climb. He enriched the fuel mixture and gave the engine more throttle; sometimes the missing stopped for several minutes. From time to time he leveled off, then started upward again until the altimeter showed 13,000 feet, comfortably above the tallest peaks. Across the Continental Divide he flew, down over the Kansas plains and into the dawn, engine running smoothly now in the warmer air. By daylight he found that he had strayed 50 miles off course—not too bad for a novice navigator. St. Louis appeared ahead and, to announce his arrival home, he buzzed Lambert Field. Down he sped at 160 miles per hour, leveled out a bare five feet above the runway, hurtled across the field and wipped into a climbing turn that carried him back to 1,500 feet. They would notice that! He landed and greeted his friends and financial backers, who had come to welcome him. Then he repaired to a nearby diner for breakfast—two huge slices of ham, four eggs, six pieces of toast. He did not mention his engine trouble.

The next morning he left for New York, and at midafternoon, after threading his way between thunderstorms—the low-pressure area had not yet moved out to sea—he landed at Curtiss Field. Photographers, alerted to the arrival of the latest Orteig Prize contender, crowded the runway, and Lindbergh had to maneuver to avoid hitting them. It was his first encounter with the world of public attention. The next few minutes amazed him. Dozens of photographers surrounded him, cursing and jostling one another and shouting rude instructions on how to pose. Reporters asked some stunningly banal questions: Did he carry a rabbit's foot? What was his favorite pie? Did he have a sweetheart?

Lindbergh found a hangar reserved for him. A team from the Wright Aeronautical Corporation awaited his pleasure: two publicity men, several engine specialists and a full-time mechanic. A fuel company representative was there, and an instrument expert, ready to install a new earth inductor compass that Lindbergh had asked for. It was an improvement over conventional compasses, which often behaved erratically under the magnetic influence of metal parts of the airplane.

Lindbergh at first felt uneasy about the publicity men, Dick Blythe and Harry Bruno, especially when he found that he was to share a room with Blythe in a hotel near Curtiss Field. As Blythe put it later, "We bedded down like two strange wildcats, each in his own hole." Soon, however, they became friends. Blythe became aware of the change in relationship at 5 a.m. one day when Lindbergh, an inveterate practical joker since his flight-school days, poured a pitcher of ice water on him.

Lindbergh never did grow accustomed to the press, however, and the strained relations that would dog him all his life began in that hectic week at Curtiss Field. An intensely private person, he gave interviews reluctantly and afterward became irritated when a flood of misinformation appeared. The newspapers nicknamed him Lucky Lindy, though those who knew him called him Slim or Charlie. On Thursday, May 19, the New York *Daily Mirror* enraged him by running a picture that showed him with a particularly fatuous grin beneath the headline: FLYIN' FOOL ADOPTS MYSTERY AIR, INDICATING QUICK TAKE-OFF. Presumably the newspaper meant the slur amicably, but Lindbergh was no fool and he was not hopping anywhere that day—the weather, in New York and over the Atlantic, remained stormy.

As coarse and inaccurate as the press coverage was, it nonetheless reflected a genuine popular fascination with Lindbergh. He was young, modest and clean-cut, and he had a winning smile. He was brave enough to challenge the Atlantic alone. And he was an outsider; no large organization backed him, no established wealth supported him. Common people everywhere admired and rooted for him, long shot that he was. The Lindbergh legend had begun.

Lindbergh's courage—foolhardiness to some—seemed magnified by the despairing search for Nungesser and Coli, who had vanished without a trace. Rumored sightings and false reports of white airplanes

Standing before the burnished cowling of his aircraft, the boyish Charles Lindbergh assumes a serious countenance for a voracious horde of press photographers after landing at Curtiss Field, Long Island.

flowed into search headquarters in New York, but nothing came of them. The loss of *L'Oiseau Blanc* emphasized the already obvious dangers of crossing the Atlantic, and opposition to such attempts began to grow. A distinguished French general, former Director of Aeronautics Marie Victor Duval, denounced all ocean flying, and *The New York Times* reported that in American government circles "Lindbergh's venture is considered suicide."

Each day Lindbergh drove with Dick Blythe to Curtiss Field. Wright technicians installed a carburetor heater on his Whirlwind engine and the earth inductor compass was added to the instrument panel. Richard Byrd offered him the use of adjoining Roosevelt Field with its mile-long runway, which Byrd had leased for his own flight in the *America*. Chamberlin and Bellanca visited, as did René Fonck and many other well-known aviation figures, men Lindbergh had never expected to meet.

Each morning and again in the evening Lindbergh checked with the weather bureau, which only the year before had started issuing aviation forecasts. Twice daily, he was told by meteorologist James H. Kimball, ships at sea and meteorological stations on land were reporting by wire and radio the barometric pressure, wind movement, temperatures, type and action of the clouds. Lindbergh liked Kimball and felt intuitively that he could trust the weatherman's judgment. Yet Kimball's estimates could be of only limited value; they were based on reports of the weather far south of Lindbergh's intended route.

Lindbergh had been in New York for a week before the weather picture began to change. On the rainy evening of May 19, Kimball reported that a high-pressure area would displace the storms from the North Atlantic and the sky should be clear over most of the route. Lindbergh had planned to attend a new Broadway musical, *Rio Rita*, but instead he headed for Curtiss Field, stopping en route at a corner drugstore to buy five sandwiches for the flight.

To Lindbergh, the time seemed ripe. He remembered his airmail rubric: Go if you can see well enough to get off the ground; you can always turn back if the weather closes in. He felt sure that if he waited until conditions were perfect, Byrd and perhaps Chamberlin, whose Bellanca had been temporarily grounded because of legal action instigated by a discharged crew member, would leave him behind. Byrd must have the same forecast, Lindbergh thought, and was probably fueling right now. But when he arrived at the field, he saw that Byrd's hangar was dark. In the morning, Lindbergh's crew would tow the *Spirit of St. Louis* to Roosevelt Field's longer runway and finish fueling it there. Then he would make the final decision on whether to go. He returned to the hotel, posted a friend outside his door to guard against interruption, and lay down at midnight to sleep.

As he tried to drift off, his mind spinning in anticipation of the next day's events, there was a knock and the door burst open. The friend posted to guard against intruders had become one himself.

"Slim," he asked, "what am I going to do when you're gone?"

The Bremen, its undercarriage and propeller smashed in an emergency landing when its fuel ran low, undergoes repairs on frigid Greenly Island off the north coast of Labrador.

A lucky landing in Labrador

On April 12, 1928, the *Bremen,* a metal-skinned Junkers monoplane, lumbered down a grass strip near Dublin, Ireland, leapfrogged a sheep and bounced into the air, bound for New York. Thirty-six hours later, the plane's tanks were almost dry and its crew—owner Guenther von Huenefeld, a German baron, and pilots Hermann Koehl of Germany and James Fitzmaurice of Ireland—had totally lost their bearings.

By sheer luck the fliers spotted a lighthouse through a gap in the clouds and managed to ditch the plane on a frozen pond nearby. The disoriented airmen learned from the lighthouse keeper that they had arrived on Greenly Island, off the northern tip of Labrador, missing New York by more than 1,000 miles. Nevertheless, they had reached North America, completing the first east-west air crossing of the Atlantic.

A German woman dressed in folk garb offers a stein of beer to the triumphant crew of the Bremen as they parade through Munich in a flower-bedecked car after their return to Europe by ship.

"I don't know," replied Lindbergh, barely hiding his irritation. "There are plenty of other problems to solve before we have to think about that one."

After that, sleep eluded Lindbergh; he simply rested, wide awake, until 2:30 a.m., then drove to his plane. Byrd's hangar remained dark, but a small crowd had already gathered at Lindbergh's hangar. His crew set out slowly in the rain, towing the plane to Roosevelt Field.

By dawn, the *Spirit of St. Louis* was poised at the west end of the runway, only a few feet from where Fonck had crashed into a gully the year before. Fully fueled, the plane weighed 5,135 pounds, slightly more than the wings were designed to lift. The five-mile-per-hour wind shifted to the west, becoming a tail wind, which would reduce lift. But Lindbergh decided not to move to the other end of the runway in order to take off into the wind because the plane would have to be towed and the tractor would cut up the clay runway, already softened by the rain.

In the closed cockpit, Lindbergh ran the engine momentarily at full throttle. His thoughts at that moment were so indelibly imprinted on his memory that, years later, he wrote of them as if they were just happening. "Thirty revolutions low! The engine's vibrating roar throbs back through the fuselage and drums heavily on taut fabric skin. I close the throttle." In the moisture-laden air, the engine had refused to run at full speed or to develop full power.

"I glance down at the wheels. They press deeply, tires bulging, into the wet, sandy clay. There is in my plane this morning more of earth and less of air than I've ever felt before."

Harry Bruno, Blythe's publicity partner, drove up in his roadster, top down despite the drizzle. A police inspector with a fire extinguisher got in. They would chase beside the plane as it headed down the runway. Anthony Fokker hurried to the far end of the field with more fire extinguishers. Two reporters were already there, huddled behind a steam roller used to pack the runway. Beyond them were telephone wires.

Inside the plane, Lindbergh made his decision.

"Do you want us to kick the blocks?" someone asked.

Lindbergh nodded. "So long," he said, and pushed the throttle forward. The great flight had begun.

Men outside heaved against the wing struts and the plane started to roll, leaving deep tracks in the soft clay. The controls felt slack in Lindbergh's palm; the plane handled like a truck, he recalled later. He watched the edge of the runway and struggled to keep the plane straight; to veer off would be fatal. The plane rolled faster; the men pushing at the struts fell behind. Lindbergh began to feel resistance in the controls as the rudder bit into the rushing air and he sensed that some of the load was beginning to shift from the wheels to the wings. As he passed the halfway mark of the runway, he realized that he was almost flying. The plane wanted to lift off but he held it to earth, straining for more speed. The cutoff point flashed by. He was committed, with not enough runway left to stop if he wanted to. Still a thousand feet from

Charles Lindbergh (right) and Clarence Chamberlin consult a map at the behest of press photographers at their hotel. Though the fliers were friendly competitors in the race to fly from New York to Paris, they were discreet about their plans and never discussed them with each other.

Just hours before Lindbergh's departure for Paris, the Spirit of St. Louis is towed tail first through the predawn drizzle from Curtiss Field to the longer runway at Roosevelt Field. "It's more like a funeral procession," thought Lindbergh, looking at the canvas shroud protecting the engine, "than the beginning of a flight to Paris."

the telephone wires, the plane took off. Lindbergh held it nearly level to gain more speed and then lifted it gently, clearing the wires by 20 feet. In the chase car below, Harry Bruno and the police inspector stared after the silver plane. Almost with one voice, they exclaimed, "By God, he's made it!" It was 7:54 a.m. on May 20, 1927.

Aloft, Lindbergh banked gingerly to avoid trees. The plane felt heavy, but the engine pulled it ever faster and the wings lifted it ever higher. Gradually, it felt more responsive, an extension of Lindbergh's own nerves and reflexes. Anxiety gone, he throttled back to 1,750 revolutions per minute, a fuel-saving pace. It was then that he first noticed other planes, hired by the press, flying alongside. Lindbergh was startled, then exasperated; he considered them intruders on his private trial.

Air turbulence often occurs where land meets water, and as the *Spirit of St. Louis* passed over Long Island Sound it began to shake, its wings flexing alarmingly. Lindbergh grew tense. The plane, after all, had never flown with such a full load. It was merely untested theory that the wings would support so much weight in rough air. But the plane absorbed every bump, and the air smoothed out.

Across the Sound, past Connecticut and over Massachusetts Lindbergh flew. Wearing a flying suit and sitting on an air cushion in a

Impressions of a solitary passage

Once he left Long Island, Lindbergh's Atlantic passage went unrecorded by camera. Ever since, artists have sought to fill the void with their impressions of his adventure—many of them inspired by the flight's 50th anniversary in 1977.

The works of five artists are reproduced on these pages. Each chose to portray a different moment and mood in the 33-hour crossing. William J. Reynolds, whose *Point of No Return* appears below, is a pilot himself who strove to create a painting that was not only beautiful but "accurate down to the smallest details, as if I had followed Lindbergh's plane across the Atlantic."

Gerald D. Coulson, *Lindbergh Crossing Atlantic*, 1976

William J. Reynolds, *The Point of No Return*, 1977

Charles H. Hubbell, *Lindbergh En Route to Paris,* 1942

Keith Ferris, *The Nineteenth Hour,* 1977

John T. McCoy, *Flying Over Valencia and Dingle Bay,* 1976

lightweight wicker seat, he felt no discomfort. The center rib of a skylight overhead had been beveled to accommodate his lanky height. Everything he needed lay within reach—a canteen of water, his sandwiches, an aluminum tube for relieving himself, the charts that would see him across the Atlantic. He left the windows on either side open.

The engine throbbed steadily and ahead the sky lightened. Bright streaks appeared as the clouds began to clear, just as Kimball had predicted. As he prepared to fly out to sea, Lindbergh put away his land maps and opened the chart on which he had laid out the great circle route to Paris. For the first hour, he would fly a heading of 63 degrees, then turn slightly south each succeeding hour.

After only three hours, the monotony of overwater flight began to grip him. He noticed details of the plane that had escaped him before. On the wing was a clump of mud, thrown up on takeoff. It irritated him unreasonably. All that effort to avoid weight and now he must carry a piece of Long Island to Paris. He ruminated on this and realized that he had fallen half asleep. His lack of sleep had already begun to affect him, and he was still 30 hours from Paris! He jerked awake.

Nova Scotia appeared on the horizon. Passing over it, he found himself only six miles off course, a two per cent navigation error; five per cent was acceptable. Suddenly local squalls engulfed him, rain drumming on fabric, but the Wright Whirlwind engine did not miss a beat. Two hours later he was over the ocean again, approaching Newfoundland. He altered course slightly to the south in order to pass above St. John's, where Alcock and Brown had started their flight eight years before. If he should disappear into the sea, at least it would be known that he had passed Newfoundland.

The need for sleep attacked him again and he fought it, forcing his eyes open, scooping cold air into the cockpit with his hand. Eleven hours out, the plane was lighter, having burned almost one fourth of its gasoline; he throttled back to 1,600 revolutions per minute, maintaining speed yet conserving fuel. It was twilight when he crossed the mountains behind St. John's, pushed over the stick and roared across the harbor. People on the ground looked up to see him pass. In a moment he was again over the vast Atlantic, North America receding, Europe still far ahead, night falling fast around him.

Stars appeared and Lindbergh used them to steer by, tilting his head back to watch them through the skylight above the cockpit. Haze began to obscure his view and he climbed from time to time to rise above it. His head began to hurt; it was pressing against the overhead rib. Looking down, he saw why; the air cushion on his seat had swollen because outside air pressure had decreased, raising him an inch or more. He let some air out of the cushion and realized for the first time that, at 10,000 feet or more, the night was bitterly cold.

Clouds that reached far higher than he could climb appeared ahead. Very well, he would plow through them. He tensed himself for flying on instruments and plunged in. Then he remembered a dangerous combi-

For 33 hours Charles Lindbergh's world was confined to the cockpit of the Spirit of St. Louis—37 inches wide, 32 inches long, 51 inches high—shown here with part of its wall fabric stripped during its restoration for public display. Seated in a wicker chair at the instrument panel, behind the main fuel tank, Lindbergh had no forward and very little side visibility.

nation: cold and moisture. He shone his flashlight outside and saw white streaks. Ice! Soon the crystallized vapor would clog the tubes that controlled his instruments and would coat the wings, weighing down the plane. Fighting the temptation to descend precipitously to a warmer altitude—if he lost control while flying on instruments he might never regain it—Lindbergh turned cautiously and groped his way out of danger. On he flew, circling the fluffy, forbidding clouds, turning southward toward the shipping lanes whenever he could not go straight ahead.

But the clouds seemed endless. "Great cliffs tower over me, ward me off with icy walls," he wrote later. "To plunge into these mountains of the heavens would be like stepping into quicksand. They enmesh intruders. They toss you in their inner turbulence, lash you with their hailstones, poison you with freezing mist." They can leave you, he added, "trying blindly to regain control of an ice-crippled airplane, climbing, stalling, diving, whipping, always downward toward the sea."

For the first time Lindbergh thought seriously of turning back. But he was almost to the halfway point, and who knew what weather lay behind him? If he did get back to New York he would have been 30 hours in the air. "Think of flying long enough to reach Ireland," he chided himself, "and ending up at Roosevelt Field!" So he pressed through the darkness, threading his plane between thunderheads.

Dawn approached. Flying into the rising sun had always made him sleepy, but now the need for sleep attacked him in ways that he had never imagined possible. He had been awake for 48 hours, flying for the last 18 of them. "Every cell of my being is on strike," he wrote, "sulking in protest, claiming that nothing, nothing in the world could be worth such effort; that man's tissue was never made for such abuse. My back is stiff; my shoulders ache; my face burns; my eyes smart. It seems impossible to go on longer." But he had no choice, *"No alternative,"* he reminded himself, *"but death and failure."*

Unstable because of its intentionally small tail surfaces, the *Spirit of St. Louis* skittered and skidded through the air. The moment the pilot's attention slackened, the airplane wandered off course. Each deviation revived him, waking him long enough to make corrections. Then he slipped away again, drifting into a hazy, hallucinatory world of half sleep. "While I'm staring at the instruments, during an unearthly age of time, the fuselage behind me becomes filled with ghostly presences. These phantoms speak with human voices, conversing and advising on my flight, discussing problems of my navigation, reassuring me."

Gradually he came to feel that mind and body had reached an impasse: Body could not stay awake; mind would not allow sleep. Somewhere between the two, he might stay alive.

Morning came, but its brightness barely roused him; still he continued to fly by instinct. Dimly he realized that when he deviated from his course, he consistently strayed to the north. The sea glinted through an opening in the clouds, and Lindbergh descended. From the spindrift whipped from the waves, he saw that he was riding a quartering tail wind that blew him relentlessly southward. Even in his sleep-waking state, he had been compensating unconsciously by steering north.

The trance continued. Island mirages appeared on the ocean ahead of him. His dulled mind suffered long periods when it seemed to sense nothing but the need to compensate for the wind's push to the south. He stopped keeping his log and lost track of his scheduled course changes. He no longer made the effort to switch between fuel tanks in the wings and fuselage to balance his plane, as he had done every hour through the night. The engine ran roughly, but he did not bother to adjust the fuel mixture. Nor could he persuade himself to figure out where he was. "Is my character so weak," wondered Lindbergh, "that I can't pull myself together long enough to lay out a new, considered course?" He slapped his face with his open hand and hardly felt the sting. His eyes fell shut and he raised the lids with his fingers. *"For the first time in my life, I doubt my ability to endure."*

Sleep, and death, seemed ready to take him. His senses swam, consciousness slipping away. He jammed his head out the window, gulping

French policemen and soldiers link hands to form a human cordon around the Spirit of St. Louis after Lindbergh's landing at Le Bourget on the night of May 21, 1927. Fearing that his plane might be dismantled by ecstatic souvenir hunters, the pilot would not leave the airdrome until he saw his faithful machine safely in a hangar.

air from the slip stream as it slashed at his face. Gradually he regained his faculties. "I've finally broken the spell," he realized. "The sight of death has drawn out the last reserves of strength."

Suddenly alert, he carefully estimated his position, accounting as best he could for all the detours around clouds, his steering errors and the effects of the wind. From his deliberations he decided that he should be flying on a heading of 115°. He looked at the compass and found that, for all his weary meandering, he was only two degrees off course.

Some hours later—he never knew precisely how many—Lindbergh saw a welcome sight below. "A boat! Several small boats. *The coast, the European coast, can't be far away!*" He dived toward one of the boats, realizing that what he hungered for, after 27 lonely hours aloft, was the sight of another human, a wave of the hand. How surprised these fishermen would be to see him. He roared over the boat but saw no one. He passed a second vessel and spied a face staring blankly at him from a porthole. He circled slowly, engine idling, just above the water.

"Which way is Ireland?" he called, but for all the reaction it showed, the face in the porthole might have belonged to a statue. Three times he circled the boat, incredulous that the sound of his engine drew no one on deck. The experience reminded him of a surreal scene from "The Rime of the Ancient Mariner," a "painted ship upon a painted ocean." Dismayed that after nearly 27 hours alone in his cockpit he was still isolated from the rest of mankind, Lindbergh reluctantly shoved the throttle forward and turned back to his southeasterly course.

At last, 16 hours out of Newfoundland, he saw land ahead, real land this time, not a mirage. Scarcely believing his eyes, he recognized Ireland's Valentia Island and Dingle Bay. After flying 3,000 miles, he was only three miles off his intended course. Soon Irishmen on the ground were waving and shouting greetings that were lost to Lindbergh in the steady throb of the Whirlwind engine.

Wide awake, Lindbergh pushed on across Ireland and over the southern tip of England to Cherbourg on the coast of France. As night fell, he followed airway beacons and the glittering River Seine until the lights of Paris rose on the horizon. To get his bearings, he circled the Eiffel Tower, which was gaudily lit with a huge electric sign advertising Citroën automobiles. Then he headed northeast in the direction of Le Bourget but could not find the airfield. He saw no beacon, no lights identifying the field, only a large black area ringed by thousands of pairs of slowly moving lights—the head lamps of swarms of cars converging on the dark space. That must be Le Bourget, he decided.

Lindbergh began a cautious descent toward a section of the unfamiliar airfield now vaguely illuminated by the automobile headlights. "Back on throttle," he instructed himself. "Bank around for final glide. The wheels touch gently. The tail skid too. Not a bad landing. I start to taxi toward the hangars, but the entire field ahead is covered with running figures!"

It had occurred to Lindbergh in his lonely cockpit that the whole world might be watching and cheering him on, and he was right. Banner newspaper headlines and radio bulletins had proclaimed that he had passed St. John's. The public had offered prayers. When his plane was sighted over Ireland, proof that he had crossed the Atlantic, a wild enthusiasm began to take hold. Now, as he landed at Le Bourget, the throng of excited Parisians awaiting him broke down a steel fence, bowled over guards and surged across the field shouting his name. Perhaps the recent, painful loss of Nungesser and Coli accounted for the magnitude of their welcome for Lindbergh. He opened the door of the *Spirit of St. Louis.*

The rambunctious crowd pulled Lindbergh from the cockpit before his feet could touch the ground and exuberantly passed him overhead from hand to hand. Someone snatched his helmet and put it on. That antic turned out to be a blessing. It decoyed the crushing mob away from Lindbergh who, with the aid of two French aviators who had come to his rescue, blended into the throng and escaped even as souvenir hunters began ripping patches of fabric from his plane. While Lindbergh hid in a room inside a darkened hangar, worrying about the plane (the damage turned out to be insignificant), the imposter wearing his helmet was introduced as Lindbergh to American Ambassador Myron T. Herrick. It was 3 a.m. before the Ambassador was able to confer quietly with the real Lindbergh in the American Embassy. The exhausted aviator consented to a short news conference with reporters who had gathered there; then, wearing pajamas borrowed from the Ambassador, he fell asleep for the first time in 63 hours.

When Lindbergh awoke around noon the next day, the embassy was aswarm with reporters. The Avenue d'Iena outside was overrun with Parisians; as the new hero stepped out on a balcony they roared a welcome. One American diplomat at first supected that the crowd was just letting off steam. "Then I realized," he said, "that Lindbergh's personality was reaching out and winning the French just as surely as his flight had reached out and found their city." In the next few days, the unassuming Midwesterner gave a series of short speeches throughout Paris. He called on Marshals Ferdinand Foch and Joseph Joffre, French heroes of World War I, and visited Nungesser's mother, graciously remarking how much more difficult her son's goal had been than his.

On May 28, his plane having been repaired at Le Bourget, Lindbergh took off for Brussels. In contrast to the French crowds, the welcoming Belgians at Evère Aerodrome were well behaved and the official ceremonies proceeded as planned. But the next afternoon, when he arrived at Croyden Aerodrome in London, the British crowd proved even more unruly than the one in Paris had been. On his first landing attempt, Lindbergh had to take off again as soon as he had touched down to avoid plowing into the mass of humanity that ran to greet him. Lindbergh spent five tumultuous days in England, during which time the *Spirit of St. Louis* was disassembled, crated and stowed aboard the cruiser U.S.S. *Memphis,* which had been dispatched from Rotterdam

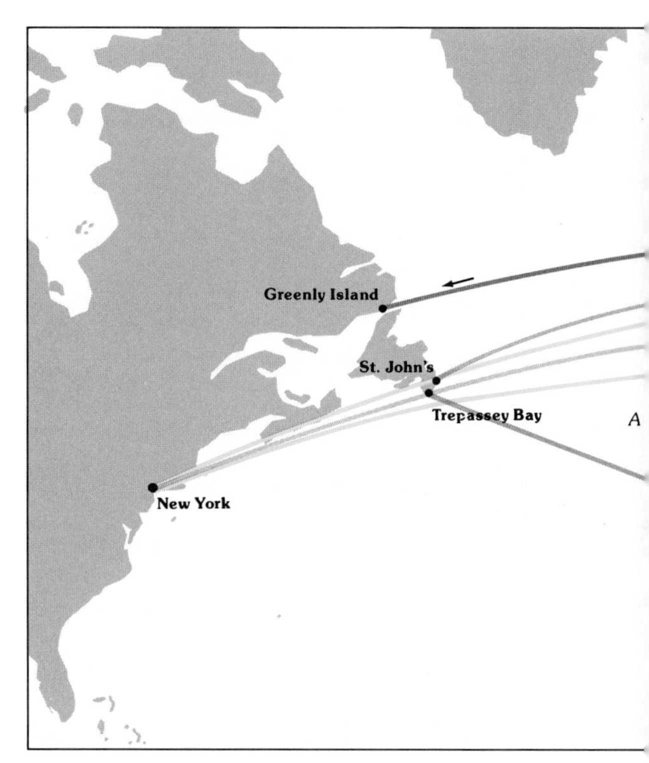

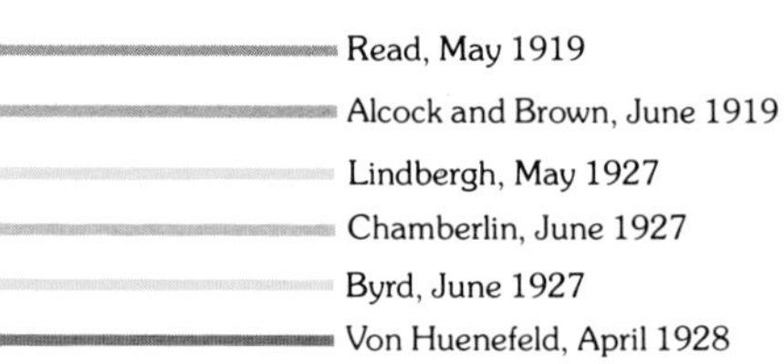

This map traces the routes of six trailblazing transatlantic flights made between 1919 and 1928. All of the pilots followed similar paths across the North Atlantic except United States Navy Lieutenant Commander Albert Read, who stopped in the Azores; all but Germany's Baron Guenther von Huenefeld flew from west to east.

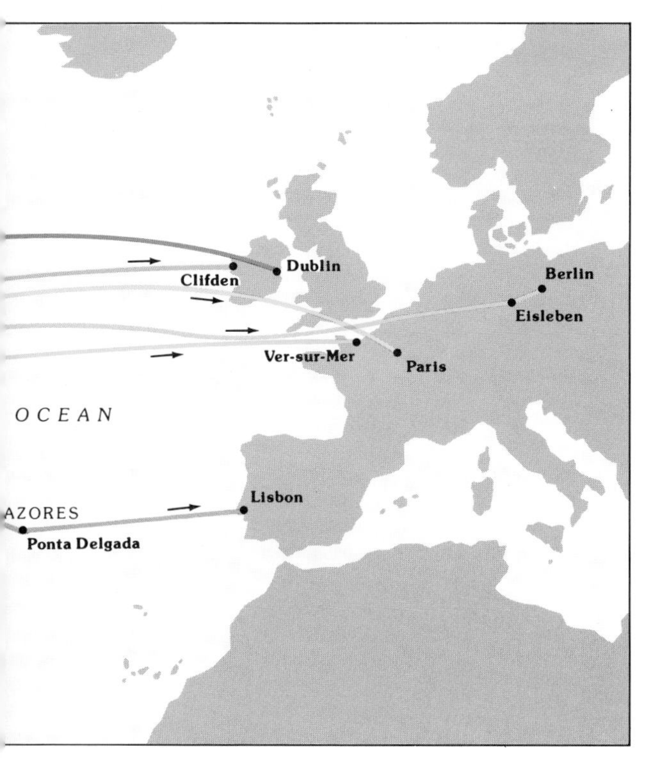

expressly to bring the hero home. Lindbergh borrowed a Royal Air Force plane and flew himself to Cherbourg, where he boarded the *Memphis.* The homecoming celebrations that awaited him in Washington and New York were triumphantly unrestrained, and on June 16, 1927, Raymond Orteig presented Lindbergh the long-sought prize, a hand-lettered check for $25,000.

In the few weeks that had passed since he left Roosevelt Field, only one of Lindbergh's competitors had made it into the air. Clarence Chamberlin, the pilot of the *Columbia,* had been in the early-morning crowd that watched the *Spirit of St. Louis* take off for Paris. The *Columbia* itself sat in a hangar guarded by two policemen; a disgruntled navigator, claiming that Charles Levine, the plane's owner, had fired him in breach of contract, had persuaded a judge to slap a temporary injunction on the entrepreneur, grounding the plane. Two days later the injunction was lifted and the Wright-Bellanca was free to fly anywhere Chamberlin or Levine wished. On receiving word that Lindbergh had reached Paris, they opted for Berlin and a new distance record. On June 4 they took off, with Levine filling in as navigator despite his almost total lack of flying experience and over his wife's horrified objections. Their plane performed as well as Lindbergh and others had been sure it would. Forty-three hours later, out of gas, they landed in a German wheat field. Their closest brush with trouble came after crossing the Atlantic when Chamberlin, after flying all night, risked a brief nap. Controlling the Bellanca at 20,000 feet was more than Levine could manage, and they went into a high-speed, spiraling dive that nearly ripped the plane apart. Chamberlin came awake and got back into his seat just in time to pull them out. Though they stopped 100 miles short of Berlin, they had flown 3,911 miles—almost 300 miles farther than Lindbergh.

On June 29, after weeks of hesitation, Richard Byrd also set off across the Atlantic, in the *America.* The damage sustained on the plane's first test flight had long since been repaired. Its crew of four included two highly respected aviators, burly Bert Acosta as pilot and Bernt Balchen of Norway as copilot. Navy Lieutenant George O. Noville handled the wireless. Byrd served as navigator and had his hands full as clouds and fog shrouded the plane for most of the journey.

The *America* reached France, but the clouds were so dense that the airmen could not find Paris. With fuel running low, Balchen took the controls from an exhausted Acosta, headed back toward the coast and ditched the *America* in the surf. Fortunately none of the crew was hurt, but no one had seen the early morning landing, so the four men paddled ashore in a rubber boat and spent the next hour trying to awaken someone to their plight.

Thus, in just a few weeks, the full stretch of the North Atlantic had been flown not once, but three times. Yet from the profusion of candidates, crashes and eventual conquests, the achievement of Lindbergh stood out like a beacon. He had done it first, and alone.

For the love of Lindy

"I was astonished," wrote Charles Lindbergh, "at the effect my successful landing in France had on the nations of the world. It was like a match lighting a bonfire." From the moment of his arrival in Paris, the hero worship and hoopla flared like fire in a high wind—threatening at times to engulf the man himself. And, unlike most such adulation, it lasted not for a few days or weeks but for years.

Handsome, modest, combining unaffected charm with unwavering determination, Lindbergh seemed to many a modern-day Galahad. In realizing his dream of flying the Atlantic alone, he had given vicarious fulfillment to the dreams of millions. In Paris, then in Brussels and London, everyday citizens cheered the young Midwesterner while royalty and heads of state paid him eloquent tribute.

President Calvin Coolidge, eager to reclaim Lindbergh as America's own, promoted him from captain to full colonel in the reserve and dispatched a Navy cruiser to bring him home to a welcome unequaled in the nation's history. In Washington, thousands cheered until they grew hoarse and applauded until their hands ached. Then Lindbergh flew to New York, where the homecoming reached a triumphant climax in a blizzard of confetti and ticker tape. At a banquet there, Charles Evans Hughes, former Secretary of State and future Chief Justice of the United States, summed up Lindbergh's impact: "We measure heroes as we do ships," he said, "by their displacement. Colonel Lindbergh has displaced everything."

At left, the Spirit of St. Louis seeks landing space at London's Croydon Aerodrome as an escort plane flies overhead. Lindbergh landed on his second try; the first was aborted when spectators swarmed onto the field. Above, Lindbergh squeezes into an open car for the ride into London.

Arriving in Washington on June 11, 1927, Lindbergh descends the gangplank of the U.S.S. Memphis. Behind him is his mother, escorted by Rear Admiral Benjamin Hutchinson and followed by the Secretaries of War and the Navy. Lindbergh said simply, "I'm genuinely glad to be back."

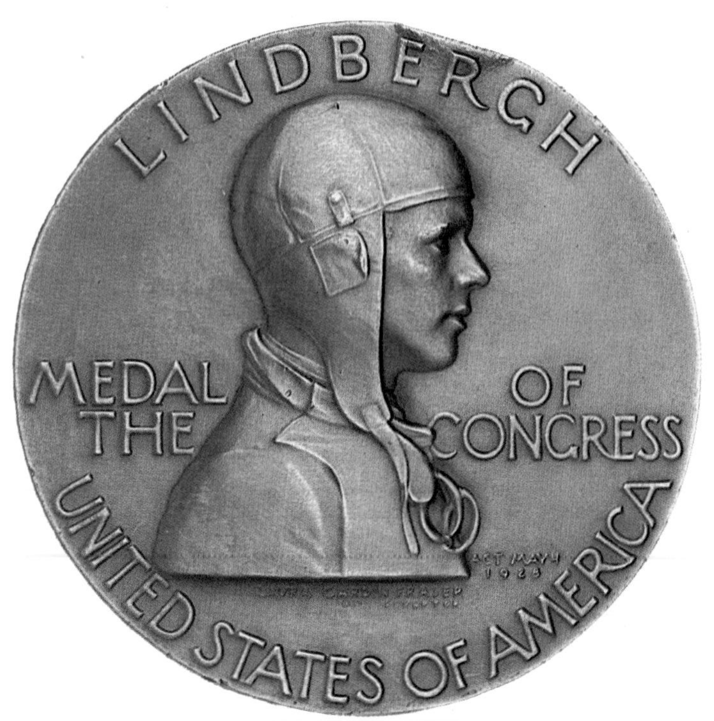

The United States Congress had this gold medal struck to commemorate the historic flight. Lindbergh became the first aviator after the Wright brothers to be so honored by vote of Congress.

President Calvin Coolidge pins the nation's first Distinguished Flying Cross on Lindbergh's lapel. The sign beneath the presidential seal refers to 48 homing pigeons, which, the Washington Evening Star stated, were released to every state in the union "to carry to the people a direct message of Lindbergh's triumphant arrival."

48 HOME COMERS!
One for every State

Lindbergh is welcomed to New York by some 500 vessels. He is on the mayor's yacht, the white-bridged boat just below the fireboat's cascade.

A delirious cascade of acclaim

"Colonel Lindbergh," declared Mayor James J. Walker, "New York City is yours. I don't give it to you; you won it." For four delirious days the city regaled Lindy like a Roman conqueror returned. In his parade up Broadway, four million New Yorkers showered him with 1,800 tons of ticker tape and confetti.

School children spelled Lindbergh's name with their bodies and civic leaders drenched him with medals and praise. On June 14, *The New York Times* devoted its first 16 pages to stories about him. And on the fourth day he collected a check for $25,000 from Raymond Orteig, the entrepreneur whose challenging offer had started it all.

NEW YORK SCHOOL CHILDREN SPELL OUT LINDBERGH

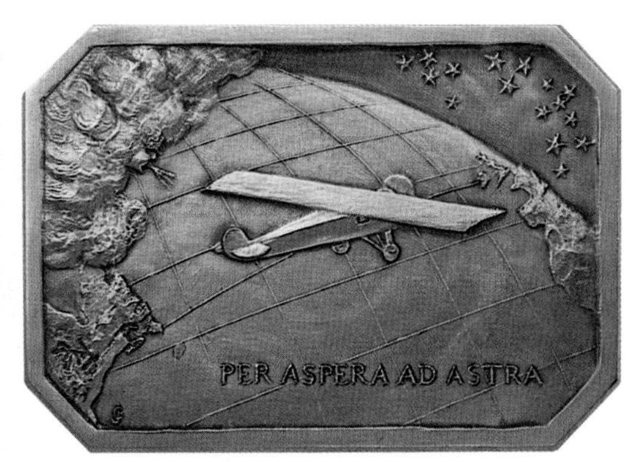

CHAMBER OF COMMERCE MEDAL

SPECIAL MEDAL OF THE CITY OF NEW YORK

MEDAL GIVEN BY BROOKLYN CHILDREN

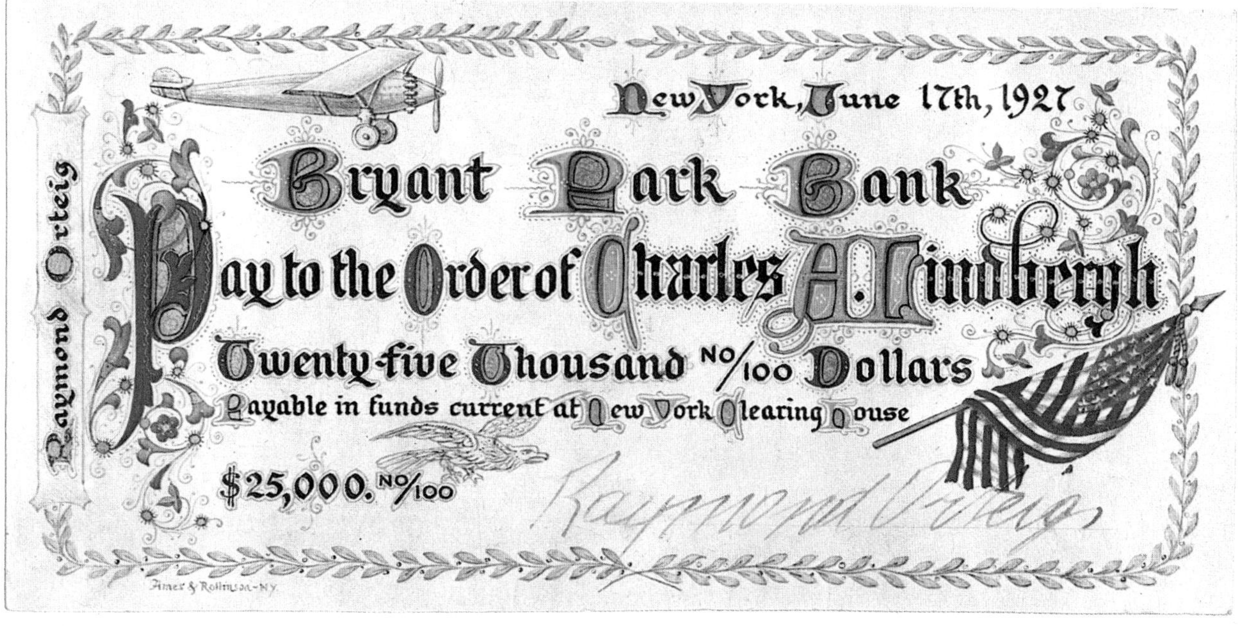
Raymond Orteig

New York, June 17th, 1927

Bryant Park Bank

Pay to the Order of Charles A. Lindbergh

Twenty-five Thousand No/100 Dollars

Payable in funds current at New York Clearing House

$25,000. No/100

Raymond Orteig

HAND-LETTERED CHECK FOR THE ORTEIG PRIZE

HAND-PAINTED SLIPPERS

PILLOW WITH LINDBERGH'S INITIALS

AIRPLANE AND EIFFEL TOWER IN A LIGHT BULB

Lindbergh was inundated with gifts—15,000 of them worth two million dollars from 69 countries. A few of the more unusual items are shown at left, above and at right. Meanwhile, businessmen sought to capitalize on Lindbergh's appeal by marketing such things as lapel buttons and music (below).

LINDBERGH LAPEL PINS

SHEET MUSIC DEDICATED TO THE HERO

A COLLAGE OF CIGAR BANDS FEATURING LINDBERGH WITH BUTTERFLY WINGS

4

The long reach to Australia

One year after the Armistice that ended World War I, four Australian veterans of that war, led by Captain Ross Smith, took off from London on a 7,388-mile odyssey home. They had adopted a motto inspired by the registration letters G-EAOU painted on the wings and fuselage of their plane: God 'Elp All of Us.

The plea for divine assistance was not lightly put. Australia, the most remote of the civilized continents, had never been reached by air. One possible approach, from the West Coast of the United States, included stretches across the vast Pacific Ocean that were simply beyond the range of any existing airplane.

An eastbound flight did appear possible—but only because most of the way was over land. The route required hopscotching across Europe and through the Middle East to India. From there the path curved along the Bay of Bengal, through Indochina and most of Indonesia, before a final leap across the Timor Sea. Anyone embarking on such a mission faced at least 20 hazardous takeoffs and landings—many of them from airstrips that hardly deserved the name. The fliers would have to carry with them the perfect selection of spare parts—too few and the plane could not be kept flying; too many and it would be overweight. Fuel had to be stockpiled at landing sites. Moreover, the airmen would fly over some of the most inhospitable deserts, jungles and mountains on earth. The peoples of some areas along the way had never seen an airplane; even among those who had, visitors fallen from the sky could not be certain of a friendly reception.

Aviators—especially Australian aviators, a notably scrappy group—could hardly ignore such a challenge. By the spring of 1919, preparations were under way for half a dozen attempts besides Ross Smith's. One of the contenders, Lieutenant Bert Hinkler, intended to be the first to do it solo. Hinkler, an Australian who had fought the War in Britain's Royal Naval Air Service, planned to use a single-engined Dove biplane supplied by the Sopwith company. He proposed to fly this lightweight aircraft to Australia in 1,000-mile hops.

In France, Étienne Poulet was motivated by entirely different circumstances. As a test pilot during the War, Poulet had put an astounding 1,400 new airplanes through their initial paces. Out of his own savings, Poulet bought a lightweight twin-engined Caudron G.4 and dedicated his flight to the memory of his close friend Jules Védrines, a record-breaking French speed and distance flier who had died in a crash early

Charles Kingsford-Smith (right) and his copilot Charles T. P. Ulm, the Australian aviators who were the first to conquer the Pacific Ocean, are memorialized in this oil painting, which hangs in the Sydney airport named for Kingsford-Smith.

in 1919. Poulet intended to donate any money he might make from the Australia venture to Védrines's widow and four children.

Before any of these missions could be launched, the concept of a flight to Australia was appropriated by the Australian Prime Minister, William Hughes, and immediately became a political issue. Hughes was in Europe to take part in the peace negotiations at Versailles, and he had shuttled between London and Paris by air. The experience convinced him that his nation, where as yet only a few civilian planes and pilots existed, needed aviation. Not only would it help tie together far-flung cities and outposts within the country but, by carrying mail and eventually passengers, it could also bring Australia closer to North America and Europe. Hughes fired off a telegram home, advocating that a prize be established for the first flight by Australians from England to Australia.

Parliament approved the plan but it was given a mixed reception by Hughes's independent-minded and somewhat isolationist countrymen, whose tax money would pay for the prize. One Australian newspaper accused the Hughes administration of underwriting "a circus flight." Another asked rhetorically, "How many people care whether there is an aerial mail between Great Britain and Australia or not?"

The debate at home did not discourage Australian aviators in Europe; the prize existed and they would try to win it. Hinkler was the first of six Australian pilots to enter the contest, which was to be run by the Royal Aero Club in London under the auspices of the Fédération Aéronautique Internationale, the authority that certified aviation records. Charles Kingsford-Smith, a former Royal Air Force lieutenant, entered soon after. With three other Australians, Kingsford-Smith planned to fly a Blackburn bomber. But the Royal Aero Club, implementing an edict of the Fédération, insisted for reasons of safety that both Hinkler and Kingsford-Smith withdraw. It was believed that Hinkler could never fly the 7,000 miles alone and that Kingsford-Smith was very unlikely to find Australia; neither he nor any of his crew knew how to navigate. Though thwarted for the moment, neither man abandoned his dream.

Among the competitors the Royal Aero Club did certify was Ross Smith, recently of the Australian Flying Corps. Smith had served with distinction in the Middle East as personal pilot to the legendary Lawrence of Arabia and had earned a chestful of decorations. Shortly after the War ended, he helped blaze an air route from Cairo to Calcutta in a Handley Page bomber. Smith appealed to Vickers Ltd. for a plane to enter the Australia race and in October 1919, Vickers turned over to him a new Vimy bomber, a sister ship to the one Vickers had provided John Alcock and Arthur Brown to fly the Atlantic a few months earlier.

By now, two other planes had started. Étienne Poulet and his mechanic, Jean Benoist, took off from Villacoublay near Paris on October 14 in the Caudron G.4, determined to reach Australia first, although as non-Australians they were ineligible for the prize. A week later, the first legitimate contestant in the race, Captain George Campbell Matthews, also with one mechanic, took off in a Sopwith Wallaby.

Three weeks later, Matthews lay snowbound in Germany, with no hope of reaching Australia within the contest's 30-day time limit. The French team had been forced down in Persia by engine trouble. While Benoist worked on the engine, hostile nomads harassed him and Poulet day and night, once throwing a torch at their aircraft. Benoist kicked the torch away before it could ignite the plane. On November 12, the engine repaired, Poulet and Benoist headed for Karachi, India (later the capital of Pakistan). The same morning, Ross Smith was ready to set out from England in late pursuit. His brother Keith, who had been a flying instructor during the War, would be his navigator; two Australian Flying Corps sergeants, Jim Bennett and Wally Shiers, who had accompanied Ross Smith on the Cairo-Calcutta flight, would go along as mechanics.

Hounslow Airdrome near London was socked in at dawn by a soupy fog. As the hours passed, the fog thinned, but visibility remained marginal. Yet Smith had to get started if he was to stand a chance of overtaking Poulet, so at 9:05 a.m., the Australians took off, on a flight

The triumphant linking of England and Australia by air is celebrated in this 1919 postcard, which traces the route of the four-man Australian crew across almost half the globe. The card was a salute from Charleville, Queensland, which the fliers visited after they reached Australia.

that would prove to be a long, suspenseful succession of small problems, serious dangers and near disasters.

Their difficulties started almost immediately. The first destination was Lyons, France, but opaque weather over the Channel forced Ross Smith to circle far wide of a direct course, with no familiar landmarks to guide him. Ice coated the wings and the fliers' goggles, nearly blinding them. "The cold is hell," wrote Ross Smith in his diary. "I am silly for having ever embarked on the flight." At last, a break in the clouds revealed a small town that the fliers were able to identify. They were only 40 miles from Lyons.

Next morning, they flew to Pisa, Italy. After they landed, a continuous downpour turned the airfield into a quagmire; for two days, takeoff was impossible. Ross Smith finally got the plane moving—although he risked leaving one of his mechanics in the mud. While he revved the engines to top power and airfield workers pushed the plane, Sergeant Bennett held down the tail to prevent the Vimy from tipping onto its nose. As the plane gathered speed for takeoff, Bennett sprinted alongside; then, with the agility of a circus performer and with a helping hand from Sergeant Shiers, he scrambled aboard at the last moment.

Miserable weather accompanied the Australians across the Apennine Mountains, where violent downdrafts—one plunged the Vimy 1,000 feet in seconds—threatened to dash the plane to earth. But between Crete and Cairo the weather improved and the crew pressed on, intent on catching up with Poulet, who had been delayed in Karachi for a week while mechanic Benoist fought off malaria.

Headed for India, the Australians encountered such strong head winds that at dusk they were forced to land at Ramadie in Iraq. The winds persisted, stirring up a violent sandstorm in the middle of the night that slammed broadside into the exposed Vimy and threatened to bowl it over. The wind snapped a control cable, allowing the ailerons to flap alarmingly. Smith ran to the cockpit and started the engines to turn the plane into the wind while 50 Indian lancers garrisoned nearby were aroused from their sleep to hold down the aircraft.

Next morning, the crew was relieved to discover that the flapping ailerons had not damaged the wings—an event that could have ended the flight—and after repairing the control cable they headed eastward again, over terrain so rough that an emergency landing would have been suicidal. Fortunately, none was necessary.

In Karachi, they learned that Poulet and Benoist were only a day's flight ahead of them. By the time the four Australians touched down in Delhi on November 25, the Frenchmen had just departed. The Indian metropolis was caught up in the thrill of the race. "One may well imagine the excitement among the residents of Delhi," commented a local newspaper, "at the departure of one Trans-Planet aeroplane in the morning and the arrival of a second one in the afternoon."

On the way to Allahabad, the next stop, the oil pressure gauge for one of the Vimy's engines suddenly dropped to zero. Without proper oil flow, the engine would seize in a matter of minutes. Ross Smith immediately landed in the desert. To his relief, the trouble proved to be only a faulty gauge, which Bennett and Shiers soon repaired. Two days later, in Akyab, Burma, the Australians at last overtook Poulet and Benoist. That night, the two crews swapped stories of their adventures and in the morning they took off in tandem for Rangoon, where the Frenchmen, in their slower plane, were left behind for good.

In Bangkok, the Australians paused for a day so Bennett and Shiers could partially overhaul the Vimy's engines. Bangkok boasted the last decent airport between London and Australia; ahead lay a chain of truly primitive landing strips. At Singora, on the Malay Peninsula, a strip had only recently been cleared of trees by workmen so unfamiliar with airplanes that they left all the tree stumps—some 18 inches tall—in the ground. Smith had no choice but to land there and thread the Vimy among the stumps. Miraculously, he missed all but one of them, which broke the tail skid—but this was one part the Vimy was not carrying among its spares. It seemed that the mission had come to an end. But the resourceful Bennett turned a new tail skid on a borrowed lathe and, after a path had been cleared through the stumps, they were ready to head for Singapore. Again, the takeoff courted disaster. Recent rains had deposited a huge puddle near the end of the airstrip, and as the Vimy plowed through the water, it slowed alarmingly. Smith was barely able to urge the plane into the sky.

The airstrip at Singapore turned out to be nothing more than the

Interlaced bamboo matting stretched over a mudbound field in Surabaya, Indonesia, forms a makeshift runway solid enough for the Smith brothers to take off on the next leg of their flight to Australia. The 900-foot strip was built in less than a day by local islanders, many of whom donated bamboo from their own homes.

straightaway of a race track. To Smith it appeared to be too short a stretch for him to stop the plane, which had no brakes. Once more he turned to Bennett. Just before the wheels touched ground, the agile mechanic climbed out of the cockpit; he inched along the bouncing fuselage toward the tail to weigh it down and force the tail skid against the ground as a brake. With Bennett clinging to the tail, the Vimy stopped just short of the curving trackside fence.

The passage to Batavia (later Djakarta, the capital of Indonesia) passed without incident. But the field at the next stop, Surabaya, proved to be a crusted-over bog into which the Vimy sank up to its axles. There seemed to be no hope of unmiring the plane for takeoff, until the Dutch governor and local Indonesians generously came to the rescue, providing hundreds of bamboo mats that they pegged to the ground as a makeshift runway 300 yards long. The light but tough bamboo supported the Vimy for the takeoff to Bima, and from there they flew easily to the island of Timor, the last stop before a 470-mile hop to Darwin, Australia, and victory. The distance would have been no obstacle for a new plane, but the Vimy had begun to show the effects of the journey; the crew feared that it might give out at any moment. Halfway to Darwin, their spirits rose at the sight of H.M.A.S. *Sydney,* an Australian warship diverted to intercept them and herald their approach.

Now they faced one last threat: The Vimy was perilously short of fuel, and Ross Smith husbanded what remained. At 3:40 p.m. on December 10, 1919, with tanks dry, he set down at Darwin's Fanny Bay airdrome. As Sergeant Shiers put it: "We almost fell into Darwin."

The trouble-plagued journey had required 28 of the 30 days allowed. Australians had been following the Vimy's progress in the newspapers for weeks. Initial cynicism forgotten, all Darwin turned out to hail the first Australians to link their homeland to Europe by air. The crowd hoisted up the crew and carried them to an impromptu reception. Knighthoods were conferred on Ross and Keith Smith; Sergeants Bennett and Shiers were promoted and awarded bars to the Air Force Medals they already had. The four men split the £10,000 prize equally.

No other contender came close. Poulet and Benoist had been stranded east of Rangoon with a cracked cylinder. Matthews and his crew, who had been snowbound in Germany when the Vimy took off, made halting progress, continuing for months after the contest was over until, on April 17, 1920, they crash-landed on the island of Bali. Of the four other planes that officially started the race, one crashed only five minutes after takeoff in England, killing the pilot and navigator; one ditched into the Mediterranean off Corfu, claiming the pilot and his mechanic; another gave up with mechanical trouble in Crete; and the last was still in England when the Smiths reached Darwin.

After the Royal Aero Club had denied Bert Hinkler permission to compete for the Australia prize, the Sopwith company had withdrawn its permission to use its plane, contending that it would be irresponsible

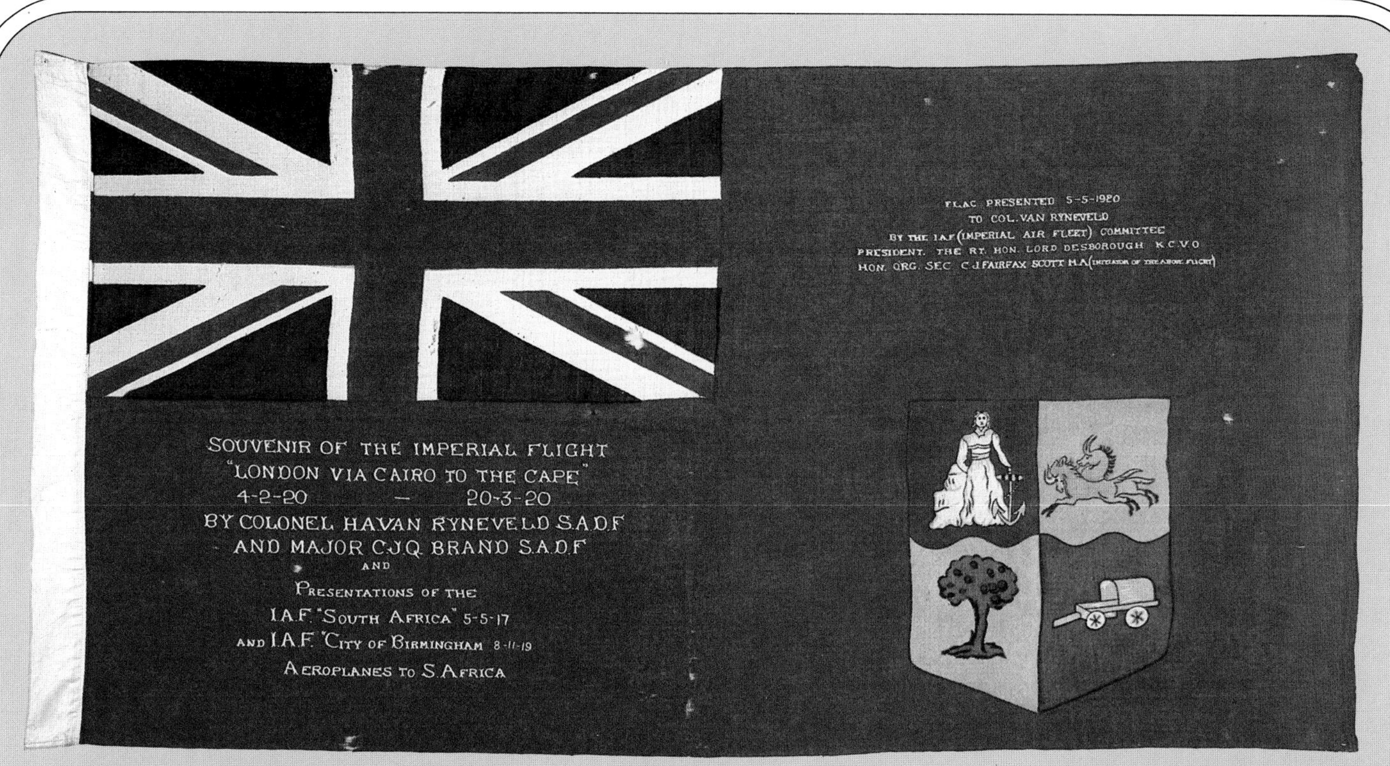

A flag dedicated to Van Ryneveld and Brand bears a Union Jack and a shield representing the four provinces of South Africa.

To Cape Town, the hard way

The excitement generated by Ross and Keith Smith's flight to Australia had barely abated when two Royal Air Force officers from South Africa took off on an adventure almost as ambitious: to fly from London to Cape Town. Lieutenant Colonel H. A. Van Ryneveld, 28, and Flight Lieutenant C. J. Q. Brand, 26, left England on February 4, 1920. Like the Smiths, they flew a twin-engined Vickers Vimy, dubbed the *Silver Queen.* But unlike the Smiths, they were embarked on a venture that seemed to be jinxed.

First the *Silver Queen's* port radiator sprang a leak and the fliers had to stay over in Italy for repairs. Then, while crossing the Mediterranean by night, the *Silver Queen* was caught in a vicious storm and the instrument panel lights failed. The fliers made it to North Africa, but as they landed in Libya, the plane's tail skid broke; they repaired it with parts from a Ford car. When radiator problems recurred over the Sahara, they made a forced landing that left the *Silver Queen* damaged beyond repair.

Van Ryneveld and Brand resumed the flight in a second Vimy, arranged for by the South African government. Still 1,200 miles from its destination, the *Silver Queen II* went down in the Rhodesian bush and was destroyed. Again, the aviators escaped injury, and again the South African government gave them a new plane—a war-surplus D.H.9. Van Ryneveld and Brand tried once more and on March 20, after 45 days and three airplanes, they became the first to reach Cape Town from London by air.

The fliers carried this medicine chest across Africa. It contained such powerful potions as lead with opium, quinine and rhubarb, and arsenic and strychnine, which, though highly poisonous, could be used in small doses to combat fatigue.

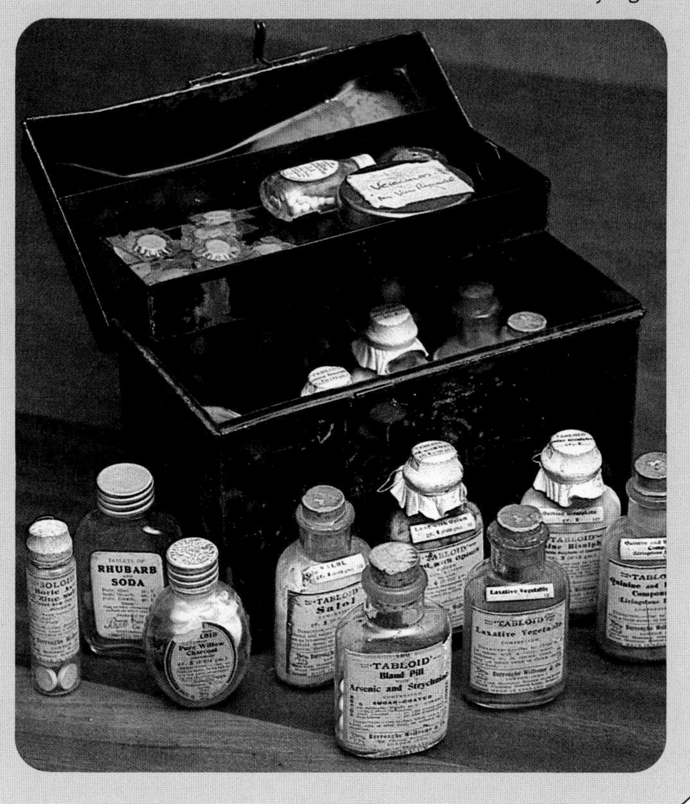

to contribute to a flight that almost surely would end in Hinkler's death.

So Hinkler bided his time. He took a job as a test pilot with A.V. Roe and Company—which allowed him to fly their airplanes to his heart's content—and saved his money to buy the beat-up prototype of the Avro Baby, a tiny plane driven by a 35-horsepower engine. Hinkler boldly decided to fly to Australia in this remarkable little plane, which at 90 miles an hour got 30 miles on a gallon of gas. After months of work to make the old plane airworthy, Hinkler took off on May 31, 1920. In three days he reached Rome, but once again his dream was thwarted; hostilities in the Middle East made passage through that region impossible. A disappointed Hinkler returned to London, crated his plane and shipped it and himself home to Australia by sea. At least he would be able to fly his Avro Baby the last 700 miles from Sydney to Bundaberg, where his parents lived.

After visiting at home for several weeks and observing firsthand the primitive state of Australian aviation, Hinkler returned to England and a job as Avro's chief test pilot. Sporting a bowler hat instead of a pilot's helmet, Hinkler earned a reputation as a crack aerobatic pilot. He married an Englishwoman and again diligently saved his money, much of it won by flying Avro planes to victory in air races.

It took four years for Hinkler to accumulate enough money to acquire an Avro Avian biplane and outfit it with an extra fuel tank for another solo try at Australia. His preparations attracted so little attention—indeed, he tried to conceal them—that when he took off from Croydon on a raw February morning in 1928, his wife, a representative from Avro and two passersby were the only spectators.

Hinkler made it to Rome in less than 13 hours—and was promptly arrested. Arriving after dark, he had mistakenly landed at a military airfield instead of a civilian one, much to the displeasure of the Italian Army. But this time Hinkler did not turn back. With assistance from the British consul, he was on his way again the next morning. He noted with some surprise, as he walked to his plane, that on his approach the night before he had unknowingly flown between radio masts hundreds of feet high.

Hinkler followed much the same route that Ross and Keith Smith had pioneered nine years earlier. It had been flown by others in the intervening years, but no one had made the flight alone. That in itself would have drawn attention to the determined Australian as he flew from country to country. But what sparked real enthusiasm was the speed at which Hinkler in his Avro was gobbling up the miles. He arrived at Basra in Iraq only five days after leaving England. "Five days ahead of Ross Smith's time," headlined one paper; at that rate, Hinkler would take only half as long to reach Australia as the Smith brothers, though his single-engined plane cruised no faster than the Vimy. The difference was Hinkler himself. A five-foot-three-and-a-half-inch dynamo, he had the stamina to see him through long days of flying and long nights of maintenance, followed by early takeoffs nearly every morning.

Shy Bert Hinkler became the object of a national craze after his 1928 solo flight from England to Australia. Australians danced the "Hinkler Quickstep," chefs concocted such dishes as "Consommé à l'Hinkler," and women of fashion took to wearing a cloche that resembled Hinkler's helmet.

As Hinkler sped across Asia, newspapers began excitedly to report his progress and he became more of a celebrity every day. Two weeks out of London, Hinkler landed near Batavia, and two days later, a little before 6 p.m. on February 22, his Avro appeared on the horizon approaching Darwin. Slowed by head winds, he was a few hours later than expected and many in the crowd gathered at the airfield that sultry summer afternoon had given up and gone home. But those who remained cheered wildly as Hinkler's plane rolled to a stop and the pilot, sunburned and unshaven, climbed out. The Mayor of Darwin greeted him, announcing to all assembled that he was "proud to shake hands with the bravest man in the world."

Hinkler's flight not only made him a popular hero but vindicated him in the eyes of aviation officialdom, which had condemned adventures like his as too risky and of no serious value. His journey had taken only 15 days, as compared with the 28 required by the Smiths. The Australian government authorized a special award of £2,000 for him; one official tribute conceded that "his was not a stunt or freak flight; it has reestablished Australia in the eyes of the world."

With Hinkler's success, the route to Australia from Europe had been explored in single-engined and multiengined planes, by multimember crews and by one man alone. But no one had yet tried to fly to Australia

Hinkler's Avro Avian is paraded jubilantly through the streets of Brisbane on the back of a truck. With its foldable wings and modest weight—910 pounds empty—the tiny Avian could be towed easily on the ground by a single person and hangared in an ordinary one-car garage.

the really hard way—across the Pacific from North America. To the uninitiated, the route appeared deceptively simple. For part of the distance, prevailing winds over the Pacific favored the flight. The course from the United States was shorter by some 1,250 miles than the route across Europe and Southern Asia. Though each of the three major segments of the transpacific route measured many more miles than the longest of Bert Hinkler's hops, none of them was quite as long as the 3,614 miles Lindbergh had flown between New York and Paris.

But Lindbergh (though his navigation turned out to be impeccable) could have erred by as much as 25° north of his course or 35° south of it and still have reached land before his fuel ran out. In the Pacific, by contrast, there was little margin for error, and therein lay the peril. If, for example, a plane flew a mere 3½° of course between California and Hawaii, the first stop on a flight to Australia, it would miss the islands entirely and must eventually fall into the endless sea.

The United States, having established a chain of military bases in the Pacific, was eager to demonstrate the feasibility of flying at least to Hawaii. The Navy had tried it first, in 1925, sending two PN-9 flying boats from San Pablo, California. Neither reached Hawaii. One was forced down after an hour by engine trouble. The second ran out of fuel a few hundred miles short of Honolulu, forcing the pilot to set down on the water; nine days later a submarine found the flying boat adrift and towed it into port. No member of either crew was seriously injured.

These early failures did not deter a young Army Air Corps lieutenant named Lester Maitland. For years he had petitioned his superiors to let him try the Hawaii flight and in 1926, when the Army purchased a Whirlwind-powered Fokker Trimotor, he asked once again. This time the Army acceded, and in February 1927, while Lindbergh was watching his plane take shape in San Diego, Maitland and his navigator, Lieutenant Albert Hegenberger, started their preparations for Hawaii.

By June the Fokker, christened *Bird of Paradise,* was ready, and on the 28th, Maitland and Hegenberger took off from the 7,000-foot strip at the municipal airport of Oakland, California, at Bay Farm Island. Minutes later they were on course for Honolulu—they hoped. In addition to the standard complement of navigation gear, the Army plane carried a revolutionary new radio designed to receive a signal beamed toward Hawaii from a transmitter, called a radio beacon, in Oakland. With this device, which had never been used before for long-distance navigation, Hegenberger hoped to be able to tell whether the plane was on course and, if not, which direction to turn to correct the error. Soon after takeoff, however, the receiver broke down. Moreover, bad weather reduced navigation to the most primitive means of dead reckoning—a standard compass and an occasional peek at the spray from the waves below to judge wind direction and speed.

Hegenberger, in Maitland's opinion, was "a navigator without peer," a man who was justifiably confident of his ability to find Hawaii. This confidence was soon put to the test. After analyzing his figures, Hegen-

A flag-waving kangaroo recites a doggerel tribute to Australia's hero of the hour, aviator Bert Hinkler, in this 1928 cartoon from the British humor magazine Punch. In fact, Hinkler's England-to-Australia flight took 15½ days, not the 16 days mentioned in the verse.

berger ordered a dramatic 17° correction in course. Since an error only one fifth that size would make them miss Hawaii completely, Maitland gulped a little as he brought the plane onto the new heading. Several hours later, aware that he had seen none of the ships they expected to pass, Maitland finally asked Hegenberger about this discrepancy. Calmly, the navigator replied that the S.S. *Sonoma,* bound from California to Hawaii, would appear in 10 minutes. Nine minutes later, Maitland saw the ship and took a deep breath of relief. Aboard the *Sonoma,* passengers rushed to the rail to catch a glimpse of the fliers, who waved as they raced by.

During the night, as Maitland climbed through clouds in the hope that Hegenberger would be able to pinpoint their position by the stars, ice blocked the carburetor of the middle engine and it quit. The Fokker sank from 11,000 feet to 3,000 before the engine fired up again on its own. Then Maitland nosed the plane slowly back to 7,000 feet—as high as it would fly without engine roughness. Hegenberger finally sighted on Polaris, the North Star, which appeared fleetingly through gaps in the clouds. The Fokker was dead on course. At dawn they saw a light ahead—Kilauea Lighthouse on Kauai, the northernmost Hawaiian island. They watched a gorgeous sunrise, then landed at Wheeler Field in a downpour, 25 hours and 50 minutes after taking off.

Two weeks after the Maitland-Hegenberger flight (and almost nine years after his disqualification from the great Australian air race of 1919), Charles Kingsford-Smith boarded the steamer *Tahiti,* bound from Sydney to San Francisco. Smithy, as everyone called him, still knew little more about navigation than he had immediately after the War; Charles Ulm, his Australian kindred spirit and newfound copilot, who sailed with him to America, was no better off. Ulm did not even have a civilian flying license, but neither man would allow that technicality, or their lack of navigational expertise, to deter them from their chosen enterprise: to fly to Australia across the Pacific Ocean.

The years since the Armistice had not been especially kind to Smithy. His first postwar venture was Kingsford-Smith, Maddocks Aeros Ltd., a company he formed in England by pooling his £150 war gratuity with those of two other Australian pilots, Cyril Maddocks and Valentine Rendle. They became the proud owners of two surplus de Havilland D.H.6s in which they toured England, offering sightseeing rides and flights to anywhere in the British Isles. But the business failed after one summer and Kingsford-Smith, despairing of persuading anyone in England to back him in a transpacific flight to Australia, decided to visit his brother Harold, who had settled in Oakland, California. There he divided his time between movie stunt-flying, barnstorming with the Moffett-Starkey Aero Circus and trying to convince any American who would listen that he could fly to Australia. Though his aerobatic daring made him useful in Hollywood and a popular attraction for the air show, he got nowhere with his plans for Australia or, because he was not a citizen,

Soldiers at Wheeler Field, Hawaii, wave to Lieutenants Lester Maitland and Albert Hegenberger as they land their United States Army transport after the successful flight from California in 1927. The 2,400-mile crossing took just under 26 hours.

with his application for a job flying the United States mail. In January 1921, he swallowed his aviator's pride and sailed home.

Thereafter, as a young barnstormer in Australia, Charles Kingsford-Smith can only be described as reckless and irresponsible. Diggers' Aviation Ltd., a company formed by ex-military pilots, hired him to hopscotch across the New South Wales countryside giving joy rides. ("Digger" was an epithet coined for Australian gold miners in the 1850s and later applied to members of Australia's armed forces.) The first time Smithy tried to land his plane he damaged the wing and undercarriage, but he had them repaired without mentioning the mishap to his employers. A few days later, while he was giving two men and a child a ride, a fractured wing spar on the plane began to buckle. He had overlooked it in his cursory examination after the accident. The plane began to spin and only Smithy's remarkable flying ability enabled him to regain control. He landed so gently that the passengers never knew of the danger they had been in.

On another day near Sydney in 1922, Kingsford-Smith zoomed four times under a bridge built only 15 feet above the Lachlan River. He

made his last pass just as two vehicles were crossing the bridge—one a wheat wagon drawn by a team of 12 horses, the other a surrey carrying a farmer and his pregnant wife to the hospital. As the plane roared under the bridge and into a loop and a half-roll overhead, the wagon team bolted. The wagon crushed the farmer's surrey against the bridge railing, tumbling the woman onto the bridge, where she bore her baby minutes later.

The same afternoon—even before the consequences of these antics caught up with him—Smithy flew two men to a neighboring ranch to help celebrate another birth. Instead of landing in an open field nearby, he set his plane down on the narrow drive to the house. The landing was unnecessarily dangerous but, except for the blowout of a tire as the plane touched the ground, Kingsford-Smith executed it perfectly. At the birthing party, however, he drank more than his share of champagne; later, while taking off, he either forgot or ignored the flat tire and veered the plane into a deep hole. He bruised his passengers, broke two of his own ribs and demolished the aircraft. When an insurance investigator discovered that Kingsford-Smith had been drinking, the company that insured the plane rejected the accident claim. This was too much even for the relatively freewheeling Diggers' Aviation company. It fired the pilot.

Smithy seemed to take the lesson to heart. Though he did not renounce his appreciation of a good belt now and then, he began to settle down in his new job as a pilot for West Australian Airways. He helped the young airline remain nearly accident-free in the perilous work of flying the mail. This record prompted his employer to remark in a letter to Smithy that "I have had excellent reports about you from all and sundry, and I am very pleased indeed with the way you are carrying on. I know now that you are all you appear to be, and this is saying a lot."

After two years with West Australian Airways, however, Kingsford-Smith and another pilot, Keith Anderson, were fired in a pay dispute. The two men abandoned aviation for a time, starting a trucking company to compete with the camel caravans that carried supplies deep into the Australian wilderness. A year later, hard work had transformed the trucking company into a prosperous if rather monotonous business, and Smithy and Anderson sold out for a profit of £2,300. It was a sizable sum, but not enough to finance a flight across the Pacific, the dream Kingsford-Smith still nurtured.

For nine years, Smithy had unsuccessfully petitioned everyone he could think of—relatives, the government, friends, even a relative of a friend—to scrape together the funds to conquer the Pacific by air. A break in this financial stalemate finally occurred after Kingsford-Smith joined forces with Ulm, also a veteran of the RAF, and one who shared his compulsion to fly the Pacific. The two men began to search in earnest for backers for the Pacific flight.

They reasoned that to start money flowing their way they needed some achievement that would catapult them into the ranks of successful long-distance fliers. To that end, they decided to fly around the perimeter of Australia, a journey of 7,500 miles, or almost the same distance as that between California and Brisbane. In 1924 the perimeter flight had been done in 22 days 11 hours—a record they set out to beat. On June 19, 1927, Kingsford-Smith and Ulm climbed into a Bristol Tourer, an outdated single-engined biplane of which Smithy was part owner, and took off from Sydney. Ten days and five and a half hours later, having cut the previous record by half, they returned to Sydney to a tumultuous welcome—and a promise from the Premier of New South Wales, J. T. Lang, of £3,500 toward their Pacific adventure.

Two weeks later Kingsford-Smith and Ulm, after persuading Smithy's former partner, Keith Anderson, to join them, set out by ship for the United States with at least enough cash in their pockets to begin organizing their epic flight. They planned to become competent navigators during the cruise north, then buy an airplane similar to the *Spirit of St. Louis,* which only six weeks earlier had carried Lindbergh from New York to Paris. But by the time they docked in San Francisco, the three men had learned enough to realize that they would need a truly experienced navigator, probably a radio operator and certainly a larger plane than Lindbergh's to carry them all.

On landing in the United States, Kingsford-Smith and Ulm were

Charles Kingsford-Smith (left) and two Royal Air Force buddies, Cyril Maddocks (center) and Valentine Rendle, tour London during World War I. After the War, the three Australian airmen started an aerial joy-riding company that went bankrupt after its two aircraft crashed.

offered a plane to fly in the Dole Race, a dash to Hawaii that was being sponsored by the pineapple company of that name. The aviators declined to participate, feeling that the offered plane was not suitable for the trip and that in any case they could never have prepared for the contest in the short time between their arrival in San Francisco and the start of the race. Their decision was a wise one. Most of the entrants were barnstormers and all of them intended to fly outmoded single-engined biplanes across more than 2,000 miles of ocean. Six of the aircraft crashed, either while preparing for the race or at the start because they could not lift off with their tremendous loads of fuel. Two others disappeared over the Pacific, and only two reached Hawaii.

These dismal statistics confirmed Kingsford-Smith's determination to assemble a qualified crew and to shop for a multiengined airplane. After much deliberation, he settled on a used Fokker Trimotor belonging to Sir Hubert Wilkins, an Australian who recently had returned from explorations in the arctic. To power the plane, which came without engines, Kingsford-Smith ordered three Wright Whirlwinds, avoiding a long delay in delivery by prevailing upon the United States government to release the engines from a production run intended for the military. The Whirlwind-powered Fokker seemed the ideal combination of plane and engines, similar planes having recently carried Commander Richard E. Byrd to Europe and Maitland and Hegenberger to Hawaii.

By early September of 1927, it appeared to Kingsford-Smith that with luck they could arrive in Australia ahead of October elections in New South Wales and thus give their sponsor, Premier Lang, a boost at the polls. But delay piled onto delay. Smithy, Ulm and Anderson, always short of cash, often were in debt even for their hotel rooms. The lack of funds slowed the preparation of the Fokker, a process that in any case took longer than expected and so discouraged Anderson that he returned to Australia. Moreover, widespread revulsion at the shocking toll taken by the Dole Race made money for a transpacific flight scarce both in Australia and in the United States. Consequently, election day came before Smithy could get started, and the result—for him as well as for his political backer—was devastating. Not only was Lang voted out, but the new government, economy-minded and determined not to be responsible for the deaths of any more fliers trying to hurdle the Pacific, ordered Kingsford-Smith to sell the Fokker and return the money to the public coffers.

Smithy was willing to convert the plane to cash, but he had no intention of giving up on the flight. He hoped to find a buyer who would then hire him and Ulm to fly the plane—locally if not to Australia—thus giving the two Australians an income while they devised another plan for tackling the Pacific. The only prospective buyer that they could find decided against the deal, and the disheartened pair had all but given up when they were introduced to Captain G. Allan Hancock, an American who owned several steamships. It was a fortuitous meeting. Fascinated by navigation and by the challenging goal that the two aviators had set

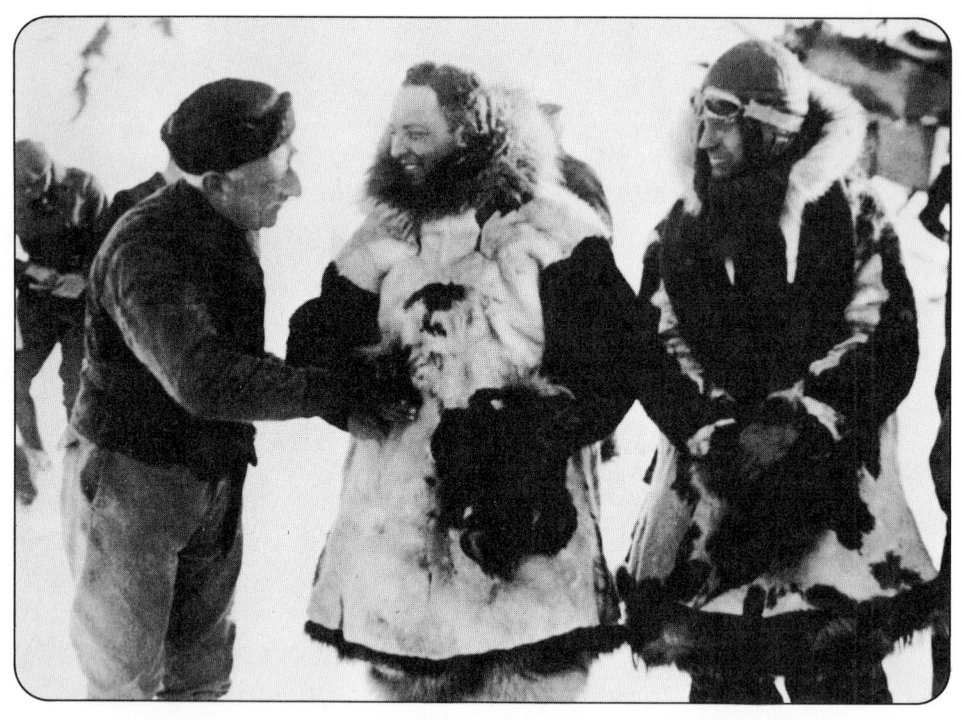

Norwegian explorer Roald Amundsen (left), who in 1911 had been the first man to stand at the South Pole, congratulates Richard Byrd (center) and pilot Floyd Bennett for making the first North Polar flight on May 9, 1926. Two days later, Amundsen followed Byrd's tracks in the dirigible Norge to become the first to visit both ends of the earth.

The Ford Tri-motor Floyd Bennett, Byrd's aircraft on the first South Polar flight, passes above the expedition's base at Little America, which he described as "a few dark roofs and a mass of trodden snow," on the rim of Antarctica. Byrd named the plane after his friend and former pilot, who had died of pneumonia in 1928.

By air to the ends of the earth

In 1929, with a 19-hour flight from his base at Little America to the South Pole and back, Commander Richard E. Byrd became the first man to have flown over both Poles. Three years earlier, with pilot Floyd Bennett, Byrd had navigated a Fokker Trimotor 1,600 miles from Spitsbergen, Norway, to the North Pole and back in 15 hours 30 minutes. (In 1959 his claim to have reached the North Pole came under severe scrutiny, but officials at the time of the flight were satisfied with Byrd's navigation reports.)

While each of the polar flights was completed in less than a day, the logistics of getting the planes and crews through treacherous pack ice to their jumping-off points—and then into the air—took months of planning and execution.

The antarctic sortie was an especially massive undertaking. Byrd put himself $184,000 in debt to outfit two ships, three airplanes and 82 men for a scientific expedition that would lock some 50 of them for two years in a frozen desert—2,000 miles from the nearest human settlement—where temperatures of –70° F. could render steel as brittle as glass. On November 28, 1929, pilot Bernt Balchen, navigator Byrd and two others flew to the Pole in just under 10 hours, jettisoning their emergency supplies on the way to gain altitude; fortunately, the return trip was uneventful.

for themselves, Captain Hancock invited them on a leisurely cruise to Mexico on his 1,400-ton yacht, *Oaxaca*. Along the way, Hancock offered not only to buy the Fokker for $16,000 but to lend the plane back to Kingsford-Smith and Ulm for their flight to Australia.

Once the *Oaxaca* returned to San Francisco, preparations for the flight advanced rapidly. The Trimotor's fuselage received a new fabric cover; emblazoned on it was a new name, *Southern Cross*, chosen for the constellation in the Southern Hemisphere that would guide the fliers home. Kingsford-Smith practiced flying the plane heavily laden with fuel. He and Ulm also practiced staying awake and active for up to 40 hours at a stretch. To complete his crew for the flight, Kingsford-Smith first tried to find a couple of qualified fellow Australians. Failing that, he chose two American merchant mariners: Harry W. Lyon, a ship's captain from Maine, as navigator; and James Warner, a radioman from Kansas City, to handle communications.

By early morning on May 31, 1928, every detail had been checked and rechecked and the big blue *Southern Cross*, ready at last, sat poised at the end of the runway at Oakland Airport. Though a curtain of mist softened the contours of the San Francisco skyline across the bay, the weather forecast promised clear skies en route to Hawaii. With a final salute to a waiting crowd, Kingsford-Smith revved up the Whirlwind engines, and the plane moved heavily along the gravel airstrip. Before it had rolled 800 feet, however, the middle engine died; Kingsford-Smith quickly cut power on the other two. As the anxious spectators clustered around the plane, navigator Lyon leaned out of the cabin, plucked a cigarette from a reporter's hand and remarked with a grin, "This may be my last drag."

Lyon need not have worried; the engine was all right. Copilot Ulm had inadvertently moved a switch that had caused excess fuel to flood the cylinders. With the switch reset, the ground crew spun the propellers again and Kingsford-Smith, without bothering to return to his starting place at the end of the runway, sent the plane trundling forward. Moments later it was in the air.

Ahead of these challengers of the earth's greatest ocean lay 2,400 miles to Honolulu and beyond that 3,200 miles to Suva in the Fiji Islands. The third and last leg of the trip—2,200 miles from Suva to Brisbane—figured to be the easiest; even an amateur navigator could hardly miss the whole of Australia.

The flight to Hawaii, it turned out, was as uneventful as a bus ride. After 27½ hours in the air, the *Southern Cross* landed at Wheeler Field with enough gas in its tanks to fly another 300 miles if that had been necessary. Only the sighting of two ships near the end of the flight had interrupted its monotony, and aside from temporary deafness caused by the incessant thunder of the three Whirlwind engines, everyone aboard, wrote Kingsford-Smith, "was chirpy as a squirrel." Even the receiver for the directional radio signals that were transmitted to guide the *Southern Cross* toward Hawaii—the same kind of apparatus that had failed Maitland and Hegenberger a year earlier—worked perfectly, giving navigator Lyon a precise check on his course for the first and last 700 miles of the trip, the limit of the transmitters' range.

The next evening, after spending a night as guests of the Royal Hawaiian Hotel and a day inspecting the *Southern Cross,* the crew ferried the plane half loaded with fuel to Barking Sands, a beach on the island of Kauai named for the sharp sound made by anyone walking on its hard-packed surface. Workmen had cleared trees from the edge of the beach to prepare a straight 4,500-foot runway—longer than the one at Wheeler Field—and had stockpiled sufficient gas to fill the Fokker's 1,304-gallon tanks to the brim. Even with a full load, Kingsford-Smith could count at best on no more than a 600-mile margin of safety on the flight to Suva. Finding the 60-by-90-mile island would be a severe test of Lyon's skill as a navigator.

Kingsford-Smith wanted above all to get an early start the next morning. He realized that whatever impetus his flight might bring to the future of air service between the United States and Australia depended almost as much on speed as on making the trip at all. He and his crew set a goal: dinner on Saturday night in Sydney. With less than a week to keep their date, Smithy warmed up the engines at 5:12 a.m., Monday, June 4, and 10 minutes later the *Southern Cross* was airborne again.

All day, Kingsford-Smith and Ulm took turns flying the plane at altitudes that rarely exceeded 800 feet. Their purpose in flying so low was to conserve fuel. They had concluded that climbing to a higher altitude with the weight of most of the fuel still aboard would burn gasoline at an extravagant rate, and that flying in thinner air required more fuel because the engines, wings and propellers worked less efficiently. But at low altitudes, the plane was at the mercy of violent squalls that soon filled the sky. Sometimes Smithy could fly around the worst of the storms; sometimes he had to plow through them. Celestial navigation was impossible, and about three hours into the flight, the radio beacon from Wheeler Field inexplicably stopped transmitting.

The rain began to leak through the seams around the cockpit windows, soaking the two pilots, and as if to heighten their discomfort, Ulm noticed an ominous dripping from the direction of fuel lines in the wing above Kingsford-Smith's head. A steady loss of fuel, even at a trickle, could dunk them into the Pacific far short of Suva. Kingsford-Smith touched his finger to the moisture, then to his tongue. The liquid was tasteless—it was only water that had condensed on the cool fuel line.

Then the right engine faltered. "There was a sputter and a sort of a kick," Ulm later reported. But the engine's vital signs held steady and soon it resumed its reassuring roar.

As night approached, Kingsford-Smith decided to expend some fuel on a steady climb upward. Daylight would last longer at a higher altitude, and if a storm came up before dawn, the extra height above the sea would offer a welcome margin of safety.

Early in the evening, storm clouds delayed the *Southern Cross* for an hour as Kingsford-Smith circled to gain enough altitude to fly over them, but as darkness deepened, the sky cleared. "Stars are popping out all over the Heavens," Ulm entered in his log. "But the Southern Cross looks best to us." The constellation for which the crew had named the plane restored their confidence that they would find Suva.

But climbing to a higher altitude apparently had cost them dearly. Smithy's calculations now convinced him that the *Southern Cross* would run out of gasoline short of land. As morning approached, and with it more storms, Ulm too turned gloomy, noting in his log at 9:30 a.m.: "Very doubtful whether we can make Suva or not."

To size up the fuel situation, the two pilots pumped the remaining gasoline into the main fuselage tank, where they could measure accurately how much was left. To their elation, the gauge showed enough for seven more hours of flight, sufficient by a slim margin to reach the island, if they were on course.

In midafternoon Suva appeared on the horizon; only minutes later Kingsford-Smith circled warily over Albert Park Sports Oval, where telegraph wires had been taken down and some trees had been felled to accommodate the first airplane ever to land in the Fiji Islands. Even with these improvements, it would be a chancy landing. The strip, only 1,300 feet long, was still bounded by trees and a fence.

A former wartime pilot who witnessed Kingsford-Smith's approach to the makeshift field at Suva from the ground found his heart in his throat. He later reported on the spectacle: "He is cutting off his engine. He is banking to land. He is coming too far in. He will have to switch on and pick up again. No! His wheels have touched. We all cry: 'He'll crash into the fence!' He's turning her. What a ground loop! He is clear of the fence, clear of the trees by a few feet. Thank God!" Ulm, much more matter-of-factly, wrote in his log: "3:50 p.m. LANDED SUVA."

Kingsford-Smith and his crew stumbled groggily from the *Southern Cross* to a rousing welcome from the Fiji Islanders. Smithy, deafened by the incessant roar of the plane's engines, could hardly understand a word spoken to him. When the Governor of Fiji tried repeatedly to invite the aviators to a ceremonial lunch, Smithy thought the man was merely trying to introduce himself.

The celebrations staged in their honor and the chore of transporting drums of fuel to a nearby island, which had a beach long enough for a fully-laden takeoff, held the *Southern Cross* on Suva for some 48 hours. With two days of hard flying ahead of them, it had become impossible for these pathfinders to keep their Saturday dinner date in Sydney, but Smithy was determined to come as close as he could.

It was midafternoon on Friday, June 8, before the *Southern Cross* lifted its wheels from the sand. Kingsford-Smith carried a good-luck whale's tooth that a Fijian elder had given him, but at first it seemed to have little effect. Shortly after takeoff, the earth inductor compass failed (the crew had forgotten to oil it before takeoff), and storm clouds began to build ahead of the plane.

The Southern Cross descends toward the rooftops of Sydney, Australia, at the end of an 8,000-mile journey that began in Oakland, California.

Kingsford-Smith tensely described the onset of the worst weather of the flight: "The visibility dwindled to a mile, then to a few yards, then to nothing. Torrential rain began to drum and rattle on the windshield. I began to climb to try to get above it. Raking gusts jolted the plane so that we had to hold onto our seats. We were tearing through a black chaos of rain and cloud at 85 knots, and our very speed increased the latent fury of the storm. We plunged on with no idea whatever of where we were. Any attempt at navigation was useless. We were circling, plunging, climbing, dodging the squalls as the poor old *Southern Cross* pitched and tossed wildly about. For four solid hours, from eight until midnight, we endured these terrible conditions."

But the plane held together; by dawn the sky had cleared and a few hours later the outline of Australia appeared on the horizon. As they sped toward land, it became clear that they had wandered off course: Moreton Island, a landmark for Brisbane, was nowhere in sight. Such a deviation approaching Suva could have been fatal, but en route to Australia the error was insignificant. Smithy corrected it by following the coast northward 110 miles to Brisbane. They were home.

In conquering the Pacific, Kingsford-Smith, with the indispensable help of Ulm, navigator Lyon and radioman Warner, had established a lasting link between Australia and its distant neighbors in North America. The welcoming crowds were impressive for a nation whose total population in 1928 was less than four million: The 15,000 spectators at Eagle Farm Aerodrome grew to thousands more along the motorcade route down Brisbane's main street. When the *Southern Cross* flew into Sydney the next day, officials estimated that 300,000 people thronged to Mascot Aerodrome there. It was Sunday, June 10. The airmen had missed their dinner date by a day, but they did not seem to care. Waiting for them was a telegram from Captain Hancock, their American benefactor, in which he offered to pay all the costs of the flight and to give the *Southern Cross,* free and clear, to Kingsford-Smith and Ulm.

The Pacific crossing launched Kingsford-Smith on a further career of pace-setting flights. And now, of course, he found backing easier to come by. In October 1930, taking off alone from London in an Avro Avian biplane, he reached Australia in 9 days 22 hours, knocking five and a half days from Bert Hinkler's record. Three years later he shortened the time again. In 1934, reversing the route that had brought him such fame in 1928, Smithy flew with Captain P. G. Taylor from Brisbane to San Francisco by way of Suva and Hawaii, becoming the first ever to fly from Australia across the Pacific to the United States.

By now, the time required for the England-to-Australia flight had been reduced by other pilots to a mere 71 hours. Determined to regain the record, Kingsford-Smith left London on November 8, 1935, flying a single-engined Lockheed Altair named the *Lady Southern Cross.* Two days later, over the Bay of Bengal, he disappeared. The only traces of him ever found were a strut and one wheel of his plane.

An aerial spectacular from Italy

In 1933, a world almost jaded by a surfeit of transoceanic flights was jolted out of its ennui by the spectacle of an entire fleet of Italian planes speeding from Italy to the Chicago World's Fair and back again. The man responsible for this achievement was Italo Balbo, Benito Mussolini's flamboyant Minister of Air, who had conceived the project to demonstrate the aerial might of Fascist Italy.

With Mussolini's blessing, Balbo spent more than a year in intensive preparation. Twenty-five Savoia-Marchetti SM.55X flying boats, each of them outfitted with two 750-horsepower Isotta-Fraschini Asso engines, had to be thoroughly tested. More than 100 pilots and crewmen had to master the tricky techniques for using water as a runway and had to become such expert formation fliers that they could keep together even through the murky weather that awaited all transatlantic aviators.

On the ground, servicing equipment had to be positioned and monumental supplies of fuel stockpiled at stopovers. To radio reports of the latest weather, a chain of six trawlers had to be strung across the Atlantic.

Balbo's armada took off from Orbetello seaplane base, some 100 miles north of Rome, on July 1. At the first stop, Amsterdam, one plane struck a dike on landing and capsized, drowning a mechanic. The remaining 24 planes left for Ireland the next morning and from there flew on to Iceland through 955 miles of rainstorms and dense fog. But the meticulous preparation paid off: The two dozen aircraft held their formation, and after landing safely at Reykjavik, continued to Chicago, arriving on July 15 after three other stops. For four days at the fair and six days in New York, celebrations of Balbo's feat continued unabated until the armada started for Rome on July 25. The fleet took a different route that included a stop in Lisbon, where a crash killed another crewman. But that final misfortune hardly dampened the tumultuous welcome when 23 of the original 25 flying boats touched down near the mouth of the Tiber River on August 12.

Balbo's remarkable accomplishment earned him not only the rank of Air Marshal of the Italian Air Force but a Distinguished Flying Cross from the United States.

Italo Balbo (third from left) inspects his twin-hulled Savoia-Marchetti SM.55X flying boat.

An endless stream of flying boats flows between Italian and American flags on this poster announcing Balbo's flight. The flight itself was part of the extended celebration—lasting for more than a year—that marked the 10th anniversary of Mussolini's rise to power in Italy.

Balbo's 25-plane fleet lines up in takeoff formation on the glassy lagoon at Orbetello seaplane base.

One of the SM.55X planes begins a test flight at Orbetello.

Balbo (right) and his copilot share a cramped cockpit in the flight.

The last planes of the armada, which flew in groups of three, cruise serenely across the Alps en route to Amsterdam.

In Chicago, the Italian flying boats lie at anchor on protected waters of Lake Michigan.

Mayor Edward J. Kelly presents Balbo with a ceremonial key to Chicago.

Italian-American women welcome a smiling aviator to the World's Fair.

Balbo sports a Sioux headdress, presented by Chief Evergreen Tree (left).

The returning flying boats sweep over a jubilant crowd of thousands gathered to greet them at Ostia, the ancient port of Rome.

Mussolini embraces Balbo in the Roman Forum during ceremonies promoting him to Air Marshal.

5

Globe-girdler from Oklahoma

Wiley Post, an unheralded Oklahoman, stands before his Lockheed Vega, the Winnie Mae, after besting several well-known airmen to win the 1930 Los Angeles-to-Chicago air derby. In the next three years, Post and the Winnie Mae would twice fly around the world, becoming the second most celebrated team in aviation—after Lindbergh and the Spirit of St. Louis.

Ignited by Lindbergh's epochal flight to Paris in the spring of 1927, the aviation world fairly exploded in a burst of other long-distance flights to every corner of the globe. The pilots were men and women of many nations. Dieudonné Costes of France, who had planned—too late—to compete for the Orteig Prize by flying from Paris to New York, found a different route to the New World. In October of 1927, with copilot Joseph Le Brix, he flew a Breguet biplane from Senegal in West Africa across the South Atlantic to Brazil. The following year two spirited British noblewomen, Mary Heath and Mary Bailey, made solo flights in opposite directions between London and South Africa, and an American, Amelia Earhart, became the first woman to cross the Atlantic, as a passenger riding with two male aviators. In 1930, Dieudonné Costes, after celebrating his South Atlantic crossing with a triumphal tour of the United States, achieved the elusive goal of the first nonstop Paris-to-New York flight.

Charles Kingsford-Smith had set a nonstop distance record on the 3,200-mile Hawaii-to-Suva leg of his transpacific flight in 1928. That mark was extended to 4,130 miles in 1929 when Squadron Leader Jones Williams and Flight Lieutenant N. H. Jenkins of the Royal Air Force flew from Cranwell, England, to Karachi, India, in 50 hours and 48 minutes. The new record stood for less than a year before the irrepressible Costes topped it with a flight of 5,000 miles from Paris to Manchuria. Rising to the European challenge, two American fliers, Russell Boardman and John Polando, flew a Bellanca from New York to Turkey, going just 12 miles farther than Costes had flown. They were lucky to make it; when Boardman switched off the Bellanca's engine in Istanbul, only a pint of fuel remained in the tanks.

The slim margin by which Boardman and Polando plucked the distance record from Costes convinced most aviators that, for the time being at least, airplanes had flown about as far as they could without landing for fuel. Multistop flights, however, no matter how great the distance, no longer seemed much of a challenge. Even to fly around the world was not quite a novelty, the feat having been accomplished in 1924 by United States Army pilots *(pages 40-41)*. Still, that journey had taken six months; if aviation were truly to shrink the globe, planes would have to be able to fly around it with much greater dispatch.

The idea of racing the clock around the world had intrigued men and women for generations. As a stunt in 1889, forty-seven years after Jules

Verne's fictional Phineas Fogg accomplished the feat in 80 days, a celebrated New York *World* reporter, Nellie Bly, actually made the trip (by train and ship) in 72 days. In 1913 a young New Yorker named John Henry Mears, who later would make his fortune as a theatrical producer, cut Nellie's time in half. But it was 1926 before anyone used an airplane in an effort to break the record. Two Americans, newspaper correspondent Linton Wells and Detroit businessman Edward S. Evans, flew whenever possible as they circled the world in less than 29 days, though they crossed both oceans by ship. Their time was soon bested by a resurgent Henry Mears. On June 29, 1928, Mears set sail for France with a folding-winged Fairchild monoplane, an airmail pilot named Charles Collyer and a Sealyham terrier mascot named Tailwind. After debarking at Cherbourg, the two men and the dog flew across Europe and Asia to Tokyo in six days. From Japan they took ship for British Columbia, then completed the circuit to New York in the monoplane. Their time: 23 days 15 hours 8 seconds.

Mears and his fellow record breakers flew primarily for adventure and had little interest in advancing aviation as a practical means of transport. One year after Mears's latest triumph, however, a German dirigible, the *Graf Zeppelin,* embarked on a world flight that its backers hoped would have a major impact on air travel. With a crew of 40 and carrying 20 passengers, who paid up to $9,000 for their tickets, the *Graf Zeppelin* departed from Lakehurst, New Jersey, on August 7, 1929. The huge airship, with accommodations as luxurious as those of any great ocean liner, crossed the Atlantic to its home base at Friedrichshafen. After refueling, it floated on to Tokyo in one long, serene swoop, arriving on August 19. From Japan, it took off for Los Angeles and then returned to Lakehurst, setting a new record of just over 21 days.

This spectacular flight, coupled with the enviable safety record that dirigibles had established, tended for a time to upstage the airplane, which, like the fabled hare, could go faster but was less likely than the tortoise to finish the course; planes crashed with dismaying frequency.

Yet to many airplane pilots, among them a one-eyed Oklahoman named Wiley Post, the apparent supremacy of the great airships was a fluke. Post understood that the essence of air travel was speed—by 1930, anyone with the $200 price of a ticket could fly in an airliner clear across the United States in 36 hours, with an overnight stop in Kansas City en route. Airships were slow. Their advantage lay in their amazing range, which had permitted the *Graf Zeppelin* to circle the entire globe with only four stops. "What I was ready and anxious to prove," Post wrote later, "was that a good airplane with average equipment and careful flying could outdo the *Graf Zeppelin.*" He was in an excellent, perhaps unique, position to do just that.

Wiley Post was born in 1898 in Grand Saline, Texas, to a large family of farmers who moved to Oklahoma when he was five. He grew up in a cornucopia; as his younger brother, Gordon, recalled, there was always

The pride of German aviation, the Graf Zeppelin, floats above Berlin's Brandenburg Gate in 1928. The mammoth airship was as long as an ocean liner—790 feet—and weighed 64 tons, but it could travel at speeds up to 70 miles per hour. In 1929 it flew around the world in 21 days.

"plenty of good food, ample comforts, lots of work." Wiley took an early interest in machines and in learning how they worked, but formal schooling was not for him and he dropped out after the eighth grade.

In 1913, taken to a county fair by his brother, Post saw his first airplane, a Curtiss pusher, and was enthralled. "To this day," he said later, "I have never seen a bit of machinery that has taken my breath away as did that old pusher." When it was nearly dark, Post confessed, "I sneaked back to where the plane was. I paced off its length and width and measured the height in 'hands' just as I had seen my father step off horses in trades. I was sitting in the rickety seat when my brother found me." Wiley returned to the farm thoroughly infected with aviation fever. He began an informal program of self-education that lasted all his life. "I read a great deal, studied mathematics by myself and experimented a bit with chemistry," he said. Post helped his family on the farm for a while and acquired some technical education at an automobile-repair school in Kansas City and an Army-sponsored radio school. Then, at 19, he found work as a roughneck in Oklahoma's flourishing oil fields. The oil boom, and the fortunes it created for freewheeling Oklahomans, were to be an impelling force in his life.

An aerobatic plane ride in 1919 proved to young Wiley's satisfaction that pilots possessed no supernatural powers, which the popular press of the day sometimes attributed to them. He saw no reason why he should not become one of them. Then one fateful afternoon in 1924, a plane passed over the drilling rig where Post was working. "The old urge to fly came over me," he recalled, "and straightway I embarked on my aviation career." A barnstorming outfit, Burrell Tibbs and his Texas Topnotch Fliers, were performing in nearby Wewoka, Oklahoma, and Post heard that Tibbs's parachute jumper had been injured. Deciding that to leap from an airplane was as good a way as any to break into flying, the 26-year-old Post went to Wewoka and asked to take the parachutist's place. Burrell Tibbs, in some doubt about Post's resolve, took him up to 2,000 feet in an old Canuck, the Canadian version of the Jenny. Post crawled onto the lower wing and buckled his harness to the chute, which was folded into a sack tied to the wing. "He then jumped," Tibbs marveled later, "as though he had done it all his life."

Post's recounting of that first jump was somewhat more dramatic. "I let go the strut," he wrote, "and backed off the wing. I swung helplessly out underneath." For several seconds Post hung there before he remembered to pull the release cord. "Suddenly all motion stopped. I seemed to be floating in the air. The plane was gone. I did not have the sensation of falling I had anticipated. With a sharp jerk, the parachute opened. I looked up and saw it spread out above me." A few seconds later, Post "swallowed hard and for the first time looked straight down. I watched the trees get bigger. Then the fences came up at me and I saw that I was going beyond the airport." He coasted over a large sycamore tree and floated softly toward a plowed meadow. As he landed, his feet caught in the furrows and pitched him onto his nose.

Little damaged, Post went on to make 99 jumps during the next two years, eventually earning as much as $200 a jump. But after a time the gloss wore off such stunts, public interest waned and the money dried up. Moreover, Post realized that to accumulate enough flying time to graduate from parachuting to piloting he would have to buy a plane of his own. So he returned to the oil fields, this time in Texas, intending to work only long enough to save the money for a plane.

On Post's first day back on the job, a roughneck on his crew chipped an iron bolt while pounding it with a sledge hammer. A sliver of metal lanced into Post's left eye, starting an infection that quickly spread and began to affect his vision in both eyes. Facing an end to his dreams of flying, Post authorized surgeons to remove his left eye. "I thought I might be blind," he said later. "When the doctor removed the bandages, one eye reacted to the lights and shadows. Little tremors ran up and down my back." His right eye had begun to mend.

Aside from the loss of peripheral vision on his left side, the worst consequence of his having only one eye was impaired depth perception. Post could not determine reliably whether one object was nearer than another or simply larger, a handicap that could be fatal to a pilot, especially on landing. "I didn't stay discouraged," he recalled. "I practiced gauging depth on hills and trees. Then I would step off the distance. At first, my mental calculations were far off, but by the end of two months I was a better judge of distance than I had ever been." Later, after he had become an accomplished pilot, Post acknowledged that "if they ever changed the height of a phone pole or a two-story building, I'd be in trouble. That's how I sight myself in."

Because of the oil-field accident, Post received a $1,800 benefit from the workman's compensation insurance program, part of which he spent to buy a damaged Canuck and have it repaired. He soon was earning a living by flying oilmen to inspect their rigs, tutoring student pilots and barnstorming on Sundays. In 1927, Post eloped with 17-year-old Mae Laine, a rancher's daughter from Sweetwater, Texas. Wiley settled Mae into the front seat of his Canuck—only to have the engine conk out a few hours into their escape. They landed in a harvested cornfield near Graham, Texas, where Wiley and Mae got the local parson to marry them.

Now that Post had a wife, he needed a steadier income than he could make as a freelance aviator. F. C. Hall, an oilman who owed his wealth to being one of the first to drill deeply—more than 7,500 feet—for oil, was looking for a personal pilot. He had just lost out on a lucrative oil lease because his competitors beat him to the bidding and he was determined never to be late again. Hall bought an airplane and Post, by virtually camping in the anteroom of Hall's offices in Chickasha, Oklahoma, got the job of flying it. But the plane had open cockpits and, as Post wrote later, "the Hall family tired of getting all dressed up like magazine aviators just to travel a few hundred miles." So Hall dispatched Post to Burbank, California, late in 1928 to pick up a new

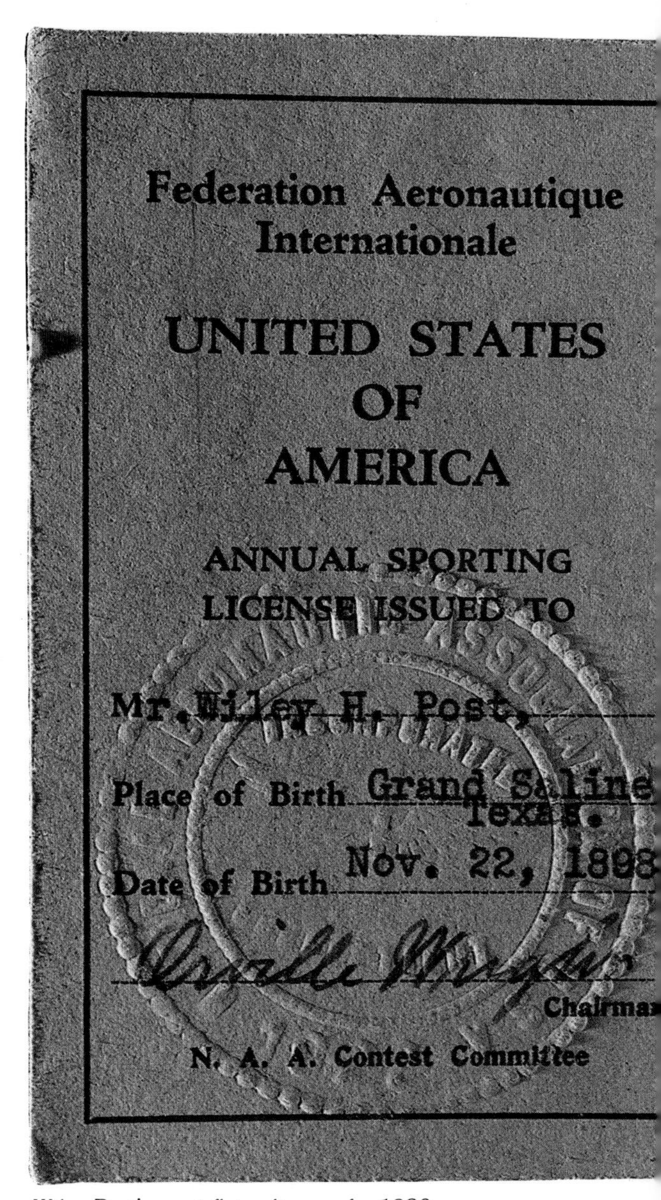

Federation Aeronautique
Internationale

UNITED STATES
OF
AMERICA

ANNUAL SPORTING
LICENSE ISSUED TO

Mr. Wiley H. Post,

Place of Birth Grand Saline
Texas.

Date of Birth Nov. 22, 1898

Chairma

N. A. A. Contest Committee

Wiley Post's sport-flying license for 1930 was signed by Chairman Orville Wright of the National Aeronautic Association, the United States affiliate of the Fédération Aéronautique Internationale, which sanctioned record-breaking attempts by accredited fliers and certified the results.

1930

License No. 576

F. A. I. Certificate No. 7547

Issued by N.A.A.

Type of Aircraft Airplane

Signature of Licensee

Valid until December 31, 1930

Lockheed Vega, which seated four in its enclosed cabin. Hall named this more comfortable aircraft the *Winnie Mae,* after his daughter.

Post suddenly found himself in enviable circumstances. The new Lockheed was as much ahead of the aviation pack as Lindbergh's Ryan had been the previous year. Moreover, his employer was an aviation enthusiast who was determined to have a role in the advance of aviation. Whenever Hall did not need his plane and pilot for business trips, Post was free to fly the *Winnie Mae* in quest of records and glory. The young aviator was full of ideas—for a record-breaking transcontinental flight or perhaps even a try at the round-the-world speed record.

But just as suddenly, Post lost his plane, and with it his job. Hall, with most of his cash invested in oil leases, reluctantly decided that Post's salary and the *Winnie Mae's* upkeep were more than he could afford. He decided to sell the plane back to Lockheed. Post had no choice but to abandon his plans half formed and ferry the *Winnie Mae* to Lockheed's used-plane lot in Burbank.

Once there, Post hired on as a salesman and test pilot for Lockheed, which at that time employed only 50 workers. For more than a year in that intimate and congenial atmosphere, he absorbed principles of drafting and aircraft design that before long would prove of immense value to him. Then, on June 5, 1930, F. C. Hall surprised Post with a telephone call from Oklahoma. Business was better and Hall wanted both Post and the *Winnie Mae* back. Told that the plane had already been sold to Nevada Airlines, Hall instructed his pilot to order another Vega. "If you can think of any improvements on a new *Winnie Mae,*" Hall's voice crackled over the long-distance wire, "go ahead and get them. I want a new ship, and I'll let you make some of those flights you were figuring on last year."

For everyday flying, the Vega needed few improvements. The fuselage was built like a streamlined barrel, two half shells of laminated plywood glued together and covered with fabric. The wings represented the latest design developed by the National Advisory Committee for Aeronautics (NACA), a government-funded research group whose work during the early 1930s helped make the United States aviation community the most innovative in the world. A new type of engine cowling, also designed at NACA to better cool and streamline radial power plants, covered a Pratt & Whitney Wasp engine of 420 horsepower. (The Wasp and the similar Wright Cyclone had deposed the Whirlwind as the best of contemporary aircraft engines.) Finally, the Vega was fast; it cruised at 150 miles per hour and had a top speed of 190 miles per hour, figures that were exceeded only by the fastest racing planes of the day.

Post asked for minor changes in the wing and tail design, then flew the new *Winnie Mae* to Oklahoma to show it off to Hall. Soon he was back in Burbank to have the plane modified for the Men's Air Derby in the 1930 National Air Races, a nonstop sprint from Los Angeles to Chicago limited to planes with a payload capacity of at least 1,000 pounds.

For the derby, Post had Lockheed install a more powerful supercharger in the Wasp engine, raising its output to 500 horsepower for a few minutes at takeoff. Applying what he had learned of aircraft design, he had the trailing edge of the wing raised to decrease drag. In place of the seats in the cabin, he asked for extra fuel tanks that would increase fuel capacity to 500 gallons. "By the time I was through testing and adjusting the ship and engine," Post wrote, "more than 10 miles an hour had been added to her top speed."

Post won the derby by more than half an hour even though a faulty compass threw him off course and added about 40 minutes to his flight time. The victory was all the more exciting because until that moment he had absolutely no national reputation as a speed pilot. The dramatic success of the *Winnie Mae* so delighted Hall that he let Post keep all of the $7,500 prize money and gave him the go-ahead for any other special flight he wished to make.

Post's dream of flying around the world was about to materialize. Within weeks, he began the myriad preparations necessary for such a flight by selecting a navigator. His choice was Harold Gatty, an Australian who had been a master navigator in the British Mercantile Marine and who subsequently had established a school of aeronautical navigation in Los Angeles. In February 1931, as Gatty attended to the details of planning a route, soliciting permission to land in all the countries along the way and arranging for supplies of gasoline to be made available for refueling, Post supervised the preparation of the *Winnie Mae* at the Lockheed factory. He ordered a snug navigator's station installed behind the cabin fuel tanks. The station had an overhead hatch for celestial and solar sightings and a seat left unattached so Gatty could slide forward or backward to compensate for the weight of fuel consumed and thus help balance the plane. An optical instrument of Gatty's invention was mounted through one side of the fuselage, enabling him to calculate at a glance ground speed and wind drift—information essential to precise navigation.

In the cockpit, Post rearranged the instrument panel to accommodate new instruments that simplified blind flying *(pages 148-149)*. He also replaced the Vega's hard steel seat with a comfortably cushioned one fitted with a folding armrest. Then, while Lockheed craftsmen tailored the plane and tuned the engine close to perfection, Post went into training for the flight. His self-designed regimen emphasized sitting—as that would be his position for many hours in the air—and breaking out of his regular sleeping cycle to stay awake long, irregular hours. He also practiced keeping his mind blank, free of extraneous thoughts and daydreams, in an attempt to turn himself into a tireless human robot at the controls.

F. C. Hall wanted the flight to begin and end at Chickasha, Oklahoma, his hometown, but up-to-date weather information for the treacherous North Atlantic crossing was impossible to obtain in the heartland of America. So Post and Gatty took off for New York, arriving at Roose-

An unheralded Pacific first

"No wheels!" The words passed apprehensively through the crowd as the red Bellanca, named *Miss Veedol,* appeared over Wenatchee, Washington, on October 5, 1931. Not only was the plane coming from the wrong direction, the east (it had taken off from Japan, far to the west, 41 hours before), but it had no landing gear.

At the controls was a flamboyant stunt pilot named Clyde Pangborn. With copilot Hugh Herndon Jr., he was about to complete the first nonstop transpacific flight, an achievement honored in Japan but little remembered in the United States—except in the town where they landed. Most of the artifacts shown here are from an exhibit in the North Central Washington Museum in Wenatchee.

Eight weeks earlier, mired in a field in eastern Siberia, Pangborn and Herndon had abandoned an attempt to set a round-the-world speed record in competition with Wiley Post. Instead they flew to Japan to pursue the $25,000 being offered by the newspaper *Asahi Shimbun* for the first nonstop flight from Japan to the United States—only to be detained for a month as suspected spies for taking photographs from the air.

By the time they were cleared to depart, Pangborn had rigged the plane's undercarriage so that it would drop off once they were airborne, thus reducing drag and extending their range for the 4,500-mile trip. But part of the landing gear did not fall off, so Pangborn climbed out on a wing strut at 14,000 feet to dislodge the remaining parts.

Fog prevented a hoped-for landing at Boise, Idaho, which would have been a new long-distance record, so Pangborn turned back to Wenatchee, where his mother lived. Thus he arrived with the morning sun, successfully bringing the *Miss Veedol* in to a spectacular belly landing *(bottom right)*.

Clyde Pangborn

Hugh Herndon Jr.

When Clyde Pangborn and Hugh Herndon (left) made the first nonstop transpacific flight in 1931 they became heroes in the Orient. The embroidered scarf below, which traces the Miss Veedol's path from Samishiro Beach, Japan, to Wenatchee, Washington, was given to Pangborn by three Shanghai school children when the pilot visited their city in 1934. Along the top border the students dedicated the scarf to "Mr. Shen Yen," a phonetic rendering of Pangborn. Along the bottom border they embroidered their own names, followed by the words "Shanghai December 1934."

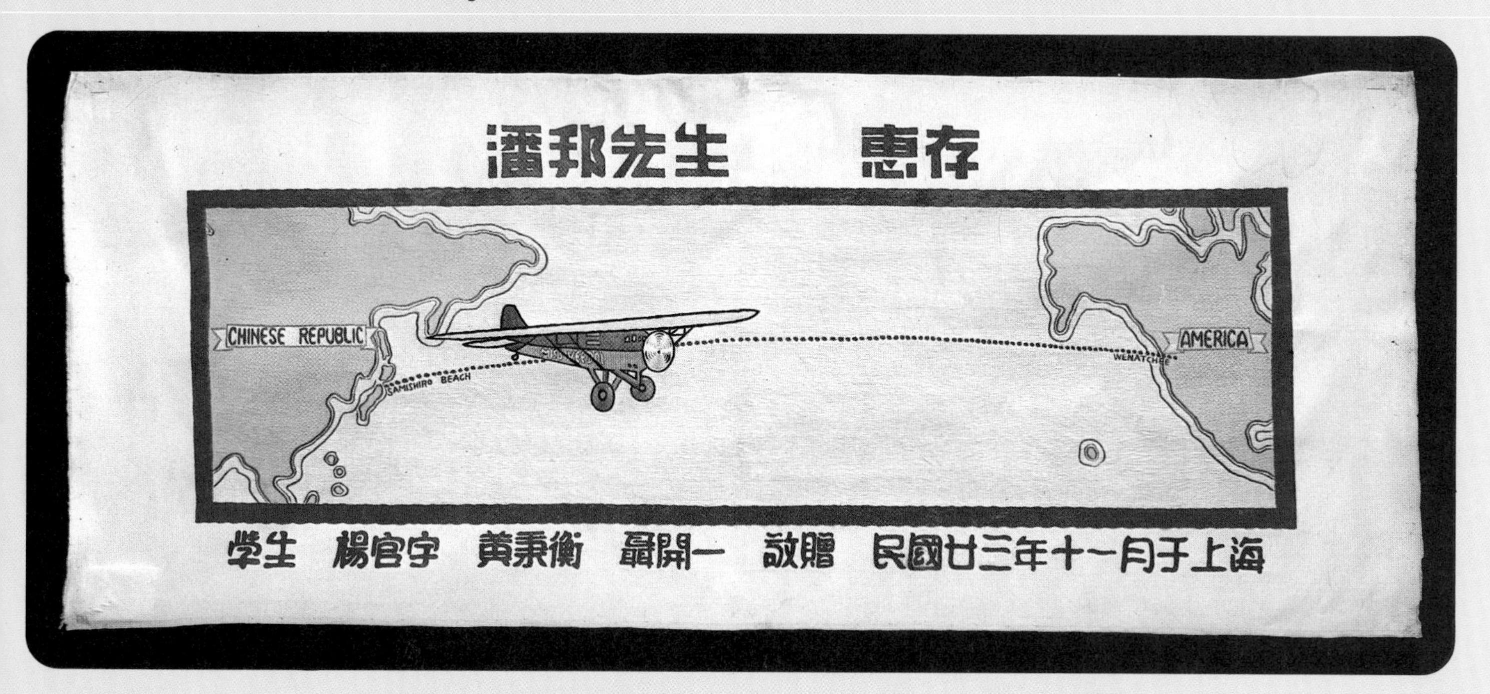

These frames from a news film of the Miss Veedol's landing at Wenatchee show the wheelless Bellanca skimming the runway at Fancher Field (left), then plowing up sand and gravel (right) as a spectator rushes forward. After a 50-foot skid on its reinforced fuselage, the plane almost nosed over when the propeller tip struck ground, but it came to rest on its belly and one wing.

velt Field on May 23, 1931. They were primed to go. But then came an exasperating delay, caused by that nemesis of all pilots, miserable weather over the North Atlantic. For 30 days storms and fog blocked the way to Newfoundland, their first stop. At last, on June 22, James Kimball, the same meteorologist who had advised Lindbergh in 1927, reported that the weather in Newfoundland was on the mend.

Post and Gatty worked through the night to prepare for takeoff. As they waited for dawn, they counted the money in their pockets—$34. But if all went well they would not need much cash; most of their expenses had been paid in advance. As the sky lightened over Long Island, Post decided to get started even though an all-night rain continued to fall. Gatty made the first entry in his log: "Took off 4:55 daylight-saving time, set course 63°, visibility poor."

Post was thoroughly at home in the cockpit, a pilot who, as a friend observed, "didn't just fly an airplane, he put it on." During the entire flight, he kept the *Winnie Mae* glued to whatever course Gatty called for, without deviation and without question—except for chiding his navigator once for specifying a course of 83½°, an impossible standard of accuracy with the instruments available. Gatty, for his part, won Post's confidence early in the flight when Woonsocket, Rhode Island, appeared below them precisely when Gatty had predicted it would.

The *Winnie Mae* touched down at Harbour Grace, Newfoundland, seven hours after taking off from Roosevelt Field. A Pratt & Whitney representative waiting at the airfield there offered to check the engine, but Post would have none of it. For one thing, he was in a hurry—he and Gatty had publicly predicted that they would complete their journey in 10 days, and privately they hoped to make it in only seven. Besides, Post insisted, the engine "was running like a watch and when you've as tough a job as this to do, it is good policy to leave well enough alone." Three hours, 43 minutes and a hot meal later, Post and Gatty were airborne for England, their load lightened by one paper dollar that Gatty had spent for sandwiches to eat en route. It was to be their only outlay of cash.

Darkness and fog forced Post to fly on instruments most of the way across the Atlantic and prevented Gatty from establishing their position with any certainty. When the navigator estimated that they should be over Ireland, Post descended gingerly through the overcast and spotted an airfield. On landing he discovered that it was the Royal Air Force's Sealand Aerodrome, near Liverpool. After lunch with the RAF, Post and Gatty flew on to Hanover, in Germany. By this time Post was so tired that he forgot to check the *Winnie Mae's* fuel gauges until he was airborne again, and to his embarrassment, had to return to Hanover for gasoline. At Berlin, the next stop, they received an enthusiastic welcome from a crowd that numbered in the thousands and were invited to dinner by German aviation officials.

Early on June 25, following a good night's sleep, Post and Gatty sped on to Moscow, a leg of nearly 1,000 miles through the worst weather

Wiley Post's Winnie Mae, its electric starter removed to save weight, is hand-started by two ground crewmen in Newfoundland on June 23, 1931. Because the plane's 420-hp engine was difficult to start, and the ground underfoot was unsure, one man swung the propeller while the other held on to brace him and keep him out of the way of the blades.

they had experienced so far; a blinding rainstorm drenched the *Winnie Mae* as if by a fire hose. Gatty purposely had Post fly north of a direct course to Moscow so that when the Volga River came into view, Gatty could be certain that a right turn would take them to their destination. The idea worked perfectly and Post landed at October Airfield at 5:40 p.m. Moscow time. No crowd greeted them there; the Soviet government had given permission through its trading agency in Washington for the *Winnie Mae* to fly across Russia. But just as the United States had not yet formally recognized the Communist government, neither did Moscow officially recognize the flight of the two Americans.

Overnight, Russian mechanics refueled the *Winnie Mae;* inadvertently, they overdid it. Post had not specified United States gallons in his instructions, so the fuel was measured in imperial gallons, which were 20 per cent larger. The resulting load was too heavy for a takeoff from the uneven surface of the Moscow airfield. Post and Gatty waited for three hours while excess fuel was siphoned from the tanks. Then, after a bumpy takeoff, strong tail winds pushed the *Winnie Mae* to Novosibirsk in central Russia at 176 miles per hour, fast enough to make up the time

The blind flight that conquered fog

Late in the morning of September 24, 1929, Lieutenant James H. Doolittle of the United States Army Air Corps sat in the rear of a two-seated training plane, his view of the outside world totally shut off by a canvas hood that covered his cockpit. The 33-year-old pilot advanced the throttle, sped along the grass and lifted into the air. His flight was the climax of an intensive year-long research program, sponsored by the Daniel Guggenheim Fund for the Promotion of Aeronautics and aimed at solving the problem of flying in cloud or fog. Guided only by a special radio receiver and the glowing dials of his instruments, Doolittle intended to fly a set course and return to his starting place—Mitchel Field, on Long Island, New York.

His eyes scanning a panel of painstakingly designed instruments *(right),* Doolittle climbed to 1,000 feet and banked to the left into a 180-degree turn. Then he straightened his plane and flew back parallel to his original course, passing beyond Mitchel Field. Turning once more, he lined up with a radio signal beamed from the ground and dropped slowly to 200 feet. He then flew level until another ground signal told him that he had neared the edge of the field below; at that point he headed for the ground. He touched down just yards from the spot where he had become airborne 15 minutes before.

In the uncovered front cockpit, Lieutenant Ben Kelsey held his hands above his head in a signal of success; he had accompanied Doolittle in case of trouble but had not once touched the dual controls. The aircraft had been flown from takeoff to landing on instruments alone.

Jimmy Doolittle's hooded flight was an extreme test, for even under the worst conditions a pilot was likely to have some visual reference to guide him—at least while he was on or near the ground. But there was no question of the test's importance: The precision instruments developed for this first public demonstration of blind flight (there had been earlier practice flights) soon became standard equipment for long-distance flying. Pilots no longer had to probe through fog or inky darkness by fallible instinct alone.

In the cockpit of his Consolidated NY-2 test plane (top), Jimmy Doolittle prepares for the trailblazing blind flight that proved the practicality of instrument flying. With the special hood in position (bottom), Doolittle was plunged into darkness; he could see nothing but the dimly lit instrument dials.

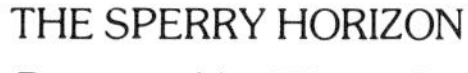
THE SPERRY HORIZON

Designed by Elmer Sperry Jr. from Doolittle's specifications, this ingenious instrument showed at a glance the aircraft's attitude in relation to the natural horizon that is a pilot's instinctive point of reference. The bar across the face of the instrument represents the horizon; the tail view of the airplane simulates the craft in flight. In level flight, the plane and the horizon bar coincide, and any change in the attitude of the plane is immediately indicated by a change in the position of the horizon bar. In the series of readings shown above (from left), the aircraft is climbing, banking to the left, banking to the right and descending. Using this artificial horizon, Doolittle was able to guide his plane precisely during a turn and keep the wings level while landing.

THE KOLLSMAN PRECISION ALTIMETER

The inner dial of this closely calibrated instrument shows a plane's altitude in thousands of feet; the outer dial registers in hundreds of feet, divided into 10-foot segments. The altitude indicated here is 264 feet. Devised by German-born engineer Paul Kollsman, this instrument responded to changes in air pressure, which diminishes as altitude increases. It was some 20 times more sensitive than the crude altimeters that were then in general use. Precise altitude readings were essential in Doolittle's 1929 blind flight, enabling him to determine the point from which he should start his glide toward the ground.

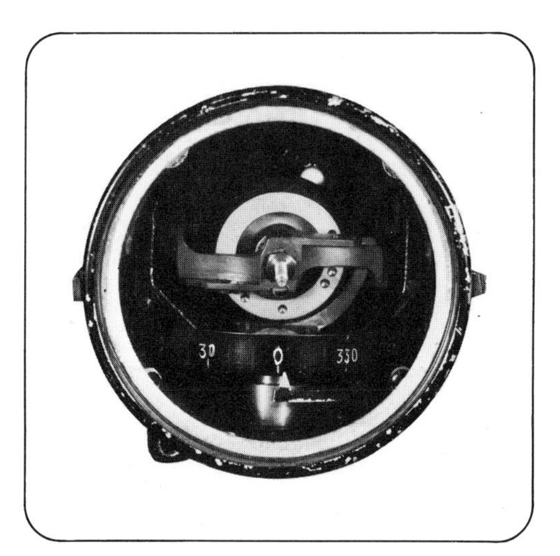

THE SPERRY DIRECTIONAL GYRO

For maintaining a straight course or gauging the amount of turn, the directional gyro at left that Elmer Sperry Jr. built for Doolittle's blind-flying project was far more reliable than the standard magnetic compass—which would swing wildly during turns or in rough weather. Calibrated in the 360 degrees of a circle, the gyro was set to correspond to the points on a compass; it would then give a steady and precise reading of any change in course. (Like the artificial horizon above, the gyrocompass had to be reset from time to time.) For his blind-flying demonstration, Doolittle plotted out in advance the desired heading for each leg of the flight.

lost in Moscow. If Post and Gatty had expected a sound night's rest in Novosibirsk, they were disappointed. Just as they were undressing for bed in the local hotel, their Russian hosts appeared and insisted that they appear at a celebratory dinner that lasted through the evening.

Post's preflight training in staying awake long hours was being put to the test. First light came early to Novosibirsk, and at 4:45 a.m. the *Winnie Mae* took off for Blagoveshchensk in Siberia, Post navigating by following railroad tracks so Gatty could nap.

After a short refueling stop at Irkutsk, they raised Blagoveshchensk just as darkness settled over Siberia. Below them, water stood two inches deep on the airfield, transforming it into a shallow lake. "We thought we were in a seaplane," wrote Post of the landing. "Spray flew all over the place. As long as the ship kept moving, we were all right, but I couldn't keep her going. I felt her left wheel sink. There I sat—in a mud hole in the last 300 miles of Siberia."

Airport officials sloshed across the quagmire in an old Ford, but efforts to dislodge the *Winnie Mae* with the automobile succeeded only in rocking the plane deeper into the muck. A tractor was summoned from a distant collective farm; before it could arrive the airfield had dried sufficiently for a team of horses to tug the *Winnie Mae,* its engine roaring exhortations to the terrified animals, out of the mud. More than 14 hours passed before Post and Gatty were airborne again. Almost five days were gone; if they had many more such delays, they would be hard pressed to finish in the predicted 10 days. But the next lap, to Khabarovsk, was accomplished without incident. A good omen, they decided.

To make up for lost time, Post decided to forgo a refueling stop at a Russian port on the Pacific Coast. Instead, he flew nonstop to Alaska, landing on the beach near the village of Solomon after 17 hours spent plowing through rain and clouds. Post prudently took on just enough fuel to reach Fairbanks, 500 miles away. But taxiing along the beach for takeoff, the *Winnie Mae* nevertheless sank into the soft sand. The plane's spinning propeller bent before Post could switch off the engine. No replacement was available but Post, the onetime oil-field handyman, managed to straighten the damaged tips of the propeller "with a wrench, a broken-handled hammer, and a round stone." Yet it seemed that the propeller would have the last word. When Gatty pulled on one of the blades to start the engine, the engine backfired and the prop smacked Gatty on the shoulder, knocking him to the ground. He suffered a wrenched back and a painful bruise—but fortunately nothing more serious—and a few minutes after the accident Post and his navigator took off.

Reporters awaited the fliers at Fairbanks and Post, though tired to the bone, promised them that in two more days the *Winnie Mae* would be in New York. Then he and Gatty slept for three hours while mechanics refilled the tanks and mounted a new propeller.

"From Fairbanks on, we began to feel the pace," wrote Gatty. The route into Canada crossed the cloud-hidden Rocky Mountains, with peaks up to 12,000 feet high. Post flew through the murk at 10,000 feet and, when Gatty told him they should have safely passed the mountains, he began a slow, cautious descent that stretched for 50 miles. The *Winnie Mae* broke out of the clouds above tracks of the Canadian National Railway, which the fliers followed into Edmonton. To their dismay, water covered the airfield there as it had at Blagoveshchensk. This time Post managed to keep his plane from sinking into the mud by landing at high speed, keeping the full weight of the *Winnie Mae* off the wheels until the last possible moment.

Getting off again, however, appeared impossible until a Canadian airmail pilot came forward with a suggestion. It might be possible to take off from Portage Avenue, a paved thoroughfare two miles long that ran from the airfield into town. Overnight, electric cables were removed from poles beside the road, yet the right of way remained alarmingly narrow. "Curbstones and electric-light poles clipped by the wing tips so fast," wrote Post of the next morning's takeoff run, "that I was just a little scared myself." Scared or not, he made a perfect takeoff. The rolling wheatlands of Alberta and Saskatchewan provinces passed uneventfully beneath them and at 5:15 p.m. Post slipped into Cleveland's Municipal Airport for a final refueling before striking out for New York. They were on the ground for half an hour, time enough to be mobbed by an enthusiastic crowd of journalists and well-wishers. Gatty tried to avoid the crush, but reporters dragged him back to face a barrage of flash cameras, ripping a pocket from his jacket in the process.

The hullabaloo in Cleveland was only a sample of what awaited the round-the-world fliers when they arrived in New York three hours later. A small fleet of planes, chartered by newspapers to carry photographers, came up to meet them, swooping aggressively around the *Winnie Mae.* Post circled the field once slowly for their benefit, then landed. An eager crowd flowed onto the field, threatening to engulf the plane, so Post cut the engine before turning off the runway and rolled to a stop—8 days 15 hours and 51 minutes after setting out. They had eclipsed the *Graf Zeppelin's* record by more than 12 days.

"When we got out, pandemonium broke loose," recalled Post of his and Gatty's emergence from the *Winnie Mae;* city police had to resort to night sticks to restore order. The excitement continued unabated for several days, through a flurry of public appearances that included a ticker-tape parade along Broadway that rivaled even Lindbergh's.

After a tour of the United States, Harold Gatty returned to his navigation school in Los Angeles. Wiley Post arranged to buy the *Winnie Mae* from Hall after the two men disagreed about Post's use of the plane for personal appearances. Then he turned his energy and the newly established power of his name to a personal campaign to raise money in order to realize another dream that he had nurtured for years: He wanted to establish the Wiley Post Institute for Aeronautical Research to advance the scientific boundaries of aviation.

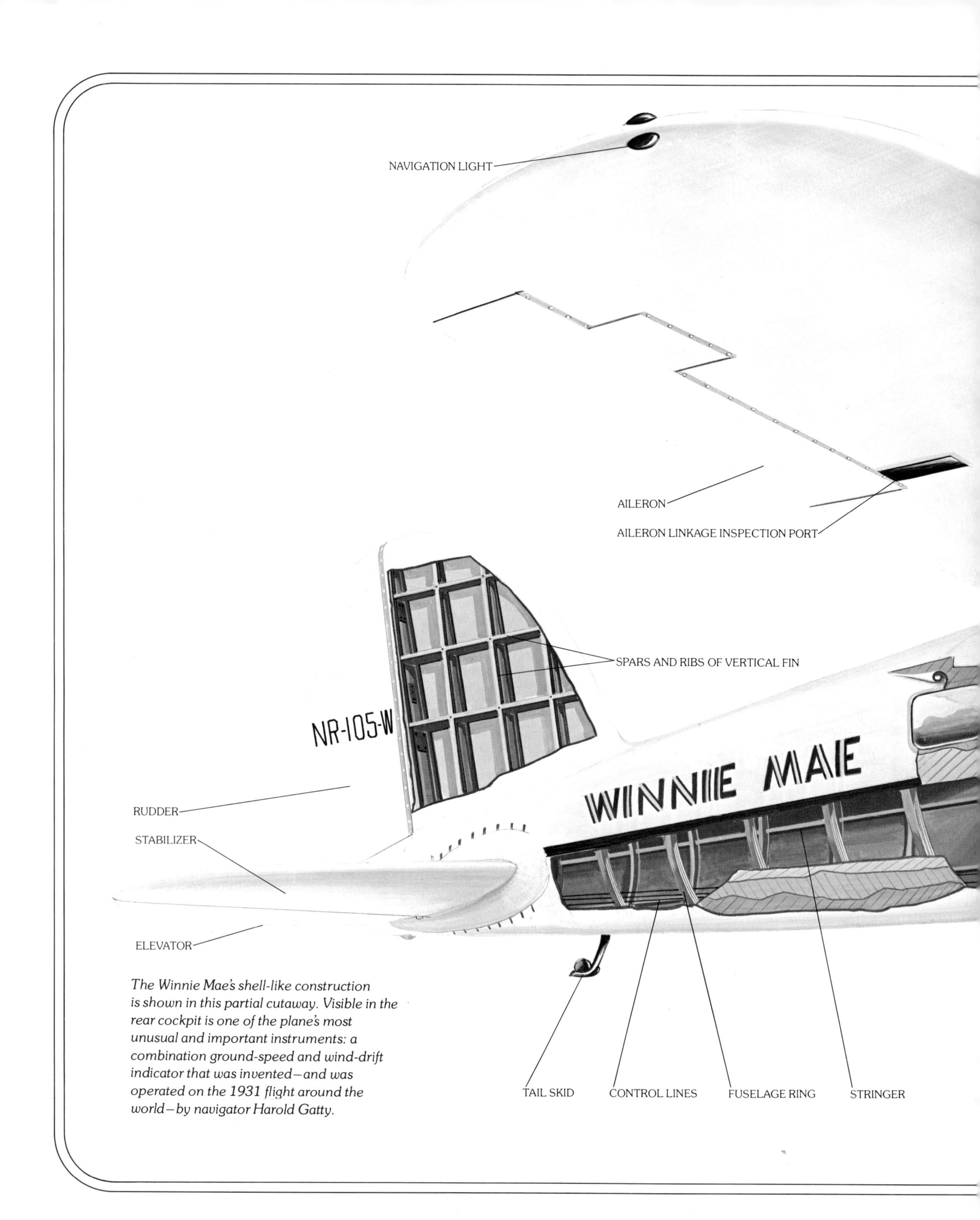

The Winnie Mae's shell-like construction is shown in this partial cutaway. Visible in the rear cockpit is one of the plane's most unusual and important instruments: a combination ground-speed and wind-drift indicator that was invented—and was operated on the 1931 flight around the world—by navigator Harold Gatty.

Inside a plywood beauty

"It was some beautiful airplane," recalled longtime Lockheed employee Harvey Christen. "There wasn't anything like it in the air." Indeed, Wiley Post's Lockheed Vega, shown here in its 1931 configuration, was far ahead of its time, a superb combination of a mighty engine—a 420-hp Pratt & Whitney Wasp—and a streamlined design.

The *Winnie Mae's* internally braced cantilever design eliminated the drag produced by struts and wires. And its lacquered plywood fuselage was an example of semimonocoque construction in which the external skin carried a large share of the stresses incurred in flight.

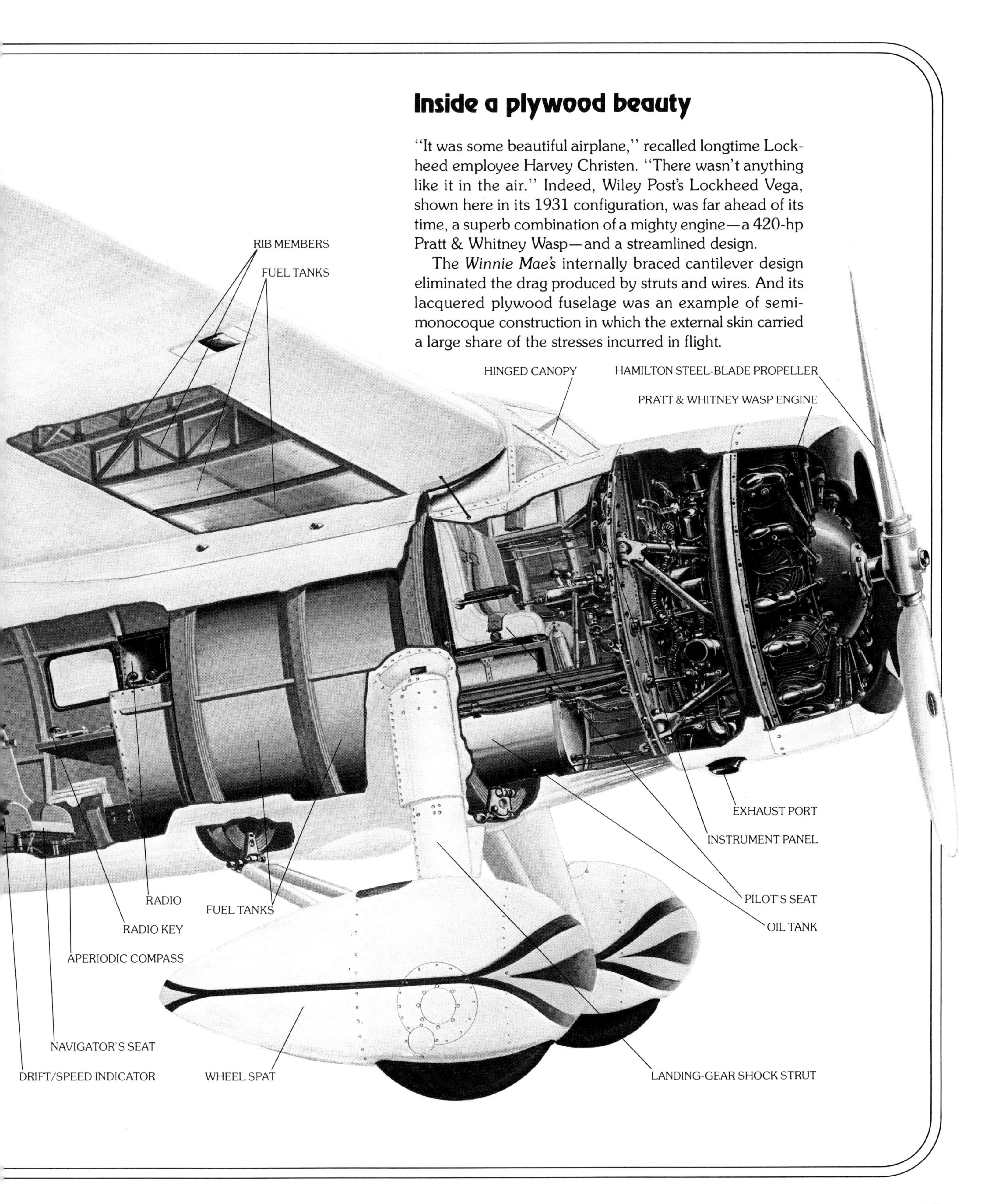

But Post could engender little support for his project. Interest in his flight gradually diminished, and funds were scarce during the early 1930s as a worldwide depression took hold. Most disappointing to Post was the realization that many people he approached felt that with his rural background and limited formal education he was unqualified to head an organization such as he planned. Some of them frankly regarded Gatty as the brains behind the world flight and Post as little more than an airborne jockey, though in fact the flight had been the result of close teamwork between them. No acrimony existed between the two men, but Post felt compelled to prove that he was as competent as Gatty. In 1933 he decided to fly around the world again—this time alone. Raising money for such a flight proved easier than raising it for aeronautical research. Post's friends in Oklahoma organized a fund-raising drive, which, along with the support of gas and oil producers and other commercial sponsors, provided all the cash he needed.

Grinning and disheveled after flying around the world in 1931, Wiley Post is flanked by his wife, Mae, and correspondent Floyd Gibbons at New York's Roosevelt Field. Asked by Gibbons to describe his record flight for the radio audience, Post said: "We had a great time."

Since 1931 there had been two major developments that made a record-breaking solo flight around the world a realistic goal. The Sperry Gyroscope Company had long been working on an autopilot system, a complex array of oil-filled hydraulic lines and automatic sensors that compensated when an airplane deviated from straight and level flight, thus relieving the pilot from having to pay incessant attention to his instruments. The first practical autopilot was not yet in production, but Post persuaded the Sperry Company to install the prototype, which he dubbed Mechanical Mike, in the *Winnie Mae*.

To simplify aerial navigation, the United States Army had perfected a radio receiver that homed in on commercial radio broadcasts. Simply by tuning in any station around the world, a pilot could use this automatic direction finder to fly toward the transmitting antenna. The Army, anxious to see how the receiver would perform over the distances Post would be flying, happily lent him one of the instruments.

As Post supervised the final details of planning the flight and preparing the *Winnie Mae*—even the running lights were removed from the wings and tail to simplify the plane's electrical system—a challenger appeared on the horizon. He was James J. Mattern, a former airline pilot who a year earlier had tried to fly a Lockheed Vega around the world; he and his navigator had reached Russia when an improperly secured hatch tore loose and the plane crashed.

In 1933, Mattern was back again in another Vega, this time to try the flight alone. However, without an autopilot or a radio direction finder, Mattern had no chance of circling the world in less time than Post could and little chance of finishing the grueling trip at all. Post refused to speed up his own preparations, even when it became clear that Mattern would take off in early June, a month before the *Winnie Mae* would be ready.

Mattern's flight was calamitous. Twice he became lost in bad weather; fatigue so overwhelmed him that he damaged his plane on nearly every landing. Each time he was able to get it repaired until, on June 14, as he neared the finish of what must have seemed a bleak and nearly endless flight across Siberia, an oil line froze, causing the engine to seize. Mattern made a belly landing on the frozen Siberian tundra, breaking an ankle and wrecking the plane.

On July 15, 1933, Post revved up the renovated *Winnie Mae* and took off for Berlin from Floyd Bennett Field, Long Island, a modern airport that boasted the world's longest concrete runways when it was opened in 1931. He intended to circle the world in just five stops. Less than 26 hours later he landed at Tempelhof Airport, becoming the first person to fly nonstop from New York to Berlin. He had navigated most of the way by dead reckoning, using the autopilot to spare himself many hours of manual blind flying. As he neared Europe, Post had enjoyed an increasing selection of radio stations for his direction finder, including one in Manchester, England, that beamed a special broadcast for him.

Because of engine vibration, joints in the hydraulic lines of the autopilot had come loose and, though an attempt to repair the device was made in Berlin, the problem became worse as Post headed for Novosibirsk, Russia, his intended second stop. At the same time, he discovered that he had left behind in Berlin maps that were crucial to his progress into Russia, and he had to land in East Prussia to replace them. Weather grounded him there overnight and in the rush the next morning, his memory failed him once more: He took off without his suitcase.

His spare clothes were expendable, but Mechanical Mike was not. Post faced a choice: Fly to Novosibirsk as planned or land first in Moscow, where he hoped to find mechanics able to repair the autopilot. Post chose Moscow, preferring to lose a little time than to fail.

The autopilot was fixed in Moscow, but the flight across the vast remainder of Russia was still difficult. Once the *Winnie Mae* almost ran into a hill that loomed suddenly out of the fog. After Novosibirsk, Post had to make another unscheduled stop in Irkutsk for more repairs to the autopilot. On the next leg he feared he would run out of fuel and

Sharing tea with his Eskimo rescuers on the eastern tip of Siberia, Texan James Mattern waits for a lift back to civilization after failing in his second attempt at a world flight in 1933. Flying solo, Mattern had crash-landed unhurt near the Anadyr River and had spent 13 days alone on the tundra before he attracted the Eskimos' attention by burning a brush-covered island in the river.

considered the prospect of bailing out. But four and a half hours after leaving Moscow, a remarkable 10 ahead of his 1931 flight with Gatty, Post landed at Khabarovsk, his jumping-off point to Alaska.

The flight to North America provided new hazards. A thick blanket of clouds covered the Sea of Okhotsk off the Russian coast, forcing Post to fly blind for several hours. But the autopilot, for a welcome change, functioned perfectly, guiding him through the murk and across the Bering Sea. At Nome, in far northwestern Alaska, his radio malfunctioned, preventing him from communicating with Fairbanks, his intended next stop, or from homing in on the city with his direction finder. As a result Post, by now thoroughly fatigued, became hopelessly lost above an overcast that concealed somewhere in its depth the highest peak in North America, 20,270-foot Mount McKinley. By pure chance, Post spotted the mining town of Flat through a chink in the overcast and landed on the crude airstrip there. He was unable to stop the *Winnie Mae* before it plunged into a ditch at the end of the 700-foot strip, crumpling its right landing gear and bending the propeller. Post himself was not injured, and damage to the plane, though disabling, was not beyond repair. Help was summoned from Fairbanks, 300 miles away, and Joe Crosson, a pilot, and Hutch Hutchinson, a mechanic, both employed by Pacific Alaska Airways, took off for Flat with a serviceable propeller. Meanwhile, men and equipment of the Flat Mining Company hoisted the *Winnie Mae* out of the ditch.

Overnight, Crosson and Hutchinson restored the *Winnie Mae* to airworthiness and Post took off for Fairbanks, following close behind the Pacific Alaska Airways' plane that had brought him help. In Fairbanks he was delayed another eight hours by bad weather, which would have made an attempt to cross the mountains into Canada foolhardy.

The trouble Post had experienced with his radio in Alaska apparently had resulted from localized weather interference, for on his next leg the static vanished and he used the direction finder to home in on Edmonton. His layover there to refuel and to replenish the oil supply in the autopilot lasted 90 minutes, then he was off to New York, the final stretch. The weather was fine and Mechanical Mike was on the job, so Post decided to doze—but not for too long at a time. He tied one end of a piece of string to his finger and the other end to a wrench, which he held in his hand. As he fell asleep, the wrench slipped from his relaxed grasp and jerked on his finger, waking him. He retrieved the wrench, checked his instruments, then nodded off again.

At Floyd Bennett Field, no fewer than 50,000 New Yorkers waited expectantly for Post to appear. It was dark and the drone of every passing airplane rekindled the crowd's excitement. Around midnight, yet another aircraft approached, but this one was different. "It has no lights!" exclaimed Post's wife Mae, who was one of those waiting. "It must be Wiley!" Floodlights drenched the runway and the *Winnie Mae* ghosted to a landing. "It was more than the crowd could bear," reported *Time*. "Thousands and thousands of excited men & women climbed

THE IRON MAN

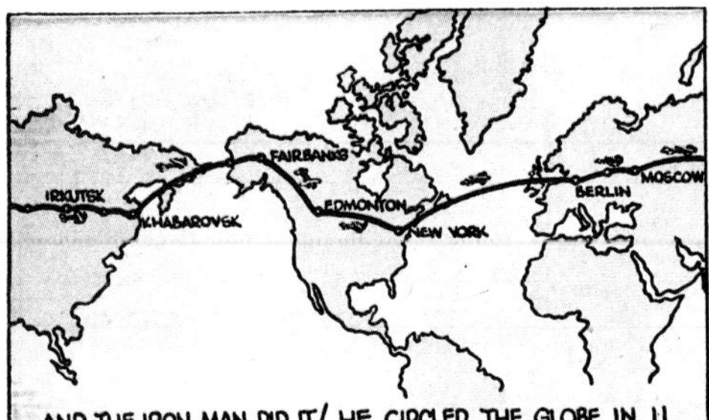

AND THE IRON MAN DID IT! HE CIRCLED THE GLOBE IN 11 HOPS FOR A NEW RECORD OF 7 DAYS, 18 HOURS AND 49½ MINUTES! HE WAS HIS OWN PILOT, HIS OWN NAVIGATOR! HE CLOCKED OFF 15,596 MILES OF FLIGHT! CONTRAST THIS WITH THE RETURN OF ONE OF MAGELLAN'S SHIPS 400 YEARS AGO

through, under and over the fences," surging toward the taxiing plane. It was precisely 11:50½ p.m., July 22, 1933. Post had flown around the world in 7 days 18 hours 49½ minutes, knocking more than 21 hours off his and Gatty's earlier record pace. *The New York Times* pointed up the true significance of Post's feat. "With the use of gyrostats and a radiocompass," said the *Times,* referring to the autopilot and direction finder, "Post definitely ushers in a new stage of long-distance aviation. The days when an almost birdlike sense of direction enabled a flier to hold his course through a starless night are over."

After the flight, Post earned several thousand dollars—but not the $50,000 his managers predicted—by promoting the oil products of his sponsors, showing newsreels of the flight at Radio City Music Hall and endorsing other products, including a leading brand of cigarettes, though he did not smoke. Post discovered that a celebrity in America sometimes faced extraordinary hazards because of the unhealthy envy of a few. As he took off after a banquet appearance in Quincy, Illinois, the *Winnie Mae's* engine died. The resulting crash sent Post to the hospital with minor injuries; more distressing was the discovery that the *Winnie Mae* had been deliberately sabotaged. Someone had poured five gallons of water into the fuel tanks.

Though Post never accumulated the money he needed to establish his institute of aeronautics, he raised enough to repair the *Winnie Mae* and later use the plane for experiments in high-altitude flying. Some experts consider this work his most important contribution to aviation.

The experiments came about when Post decided to enter the MacRobertson Race from England to Australia, set for late 1934 to coincide with the 100th anniversary of Melbourne's founding. The *Winnie Mae* was fast becoming obsolete and Post realized that to win he would have to fly at 30,000 feet or more. At that altitude the *Winnie Mae,* equipped with a variable pitch propeller, would be faster than the competition flying in the denser air closer to the earth. Moreover, Post's studies convinced him that at high altitude he would be helped by westerly tail winds of 100 miles or more per hour. All he lacked was a breathing apparatus suited to the rarefied atmosphere six miles above sea level.

Though the world altitude record in 1934 stood at 47,352 feet, the pilot who set it had remained there for no more than a few minutes; oxygen equipment of the time was inadequate to sustain a pilot for much longer at such a height. Post knew that the long-range solution was a pressurized cabin, but the plywood *Winnie Mae* could not be made airtight, so he set out to solve the problem by inventing a pressure suit that would make him feel as if he were flying at 5,500 feet regardless of the altitude. His first design, built by the B. F. Goodrich Rubber Company, blew out a seam when it was tested at Wright Field in Ohio, where the Army maintained an altitude chamber. The second suit fit Post so tightly it had to be cut off him. But the third model was a success and he used it on September 5, 1934, to climb comfortably to nearly

Wiley Post's status as a national hero is reflected in this syndicated comic strip of 1934, which traces the tale of his life from Oklahoma boyhood through two world flights and compares him to Magellan.

40,000 feet. Though the suit performed well, the test flight exposed some mechanical deficiencies in the *Winnie Mae.* By the time they were corrected, the MacRobertson Race had been won by someone else.

Nevertheless Post remained determined to prove that flying at high altitude was the key to high-speed air transportation. His next target was the transcontinental speed record. Four times in a period of five months he tried for it and four times, for various reasons, he failed. He ran short of fuel or oxygen, or the *Winnie Mae's* supercharger failed. Once a rival pilot enlisted a mechanic to sabotage the plane. Finally, the *Winnie Mae's* trusty Wasp engine simply wore out.

But Post proved his main point. On one of his transcontinental attempts, pushed along by strong tail winds, the *Winnie Mae* reached a speed of 340 miles per hour, almost twice as fast as the average speed of the plane that won the MacRobertson Race.

In June of 1935 an Oklahoma congressman introduced legislation under which the government would buy the venerable *Winnie Mae* for $25,000 and display it at the Smithsonian Institution in Washington. Anticipating the sale, Post bought a new airplane, a hybrid constructed from two planes that had been damaged: the fuselage of a Lockheed Orion and the wings of a Lockheed Explorer.

After his first world flight, Post had met Will Rogers, the immensely popular American folk humorist, and in summer of 1935, Rogers hired Post to fly him to Alaska in search of new material for his syndicated newspaper column. In particular he intended to interview the "King of the Arctic," a whaler and trader named Charlie Brower who had lived for 50 years near Point Barrow on the northernmost edge of Alaska.

Post had ordered pontoons to convert his Orion-Explorer to a seaplane for landing on Alaska's lakes, but they were late arriving in Seattle, where the expedition was to begin. Rogers was impatient to be off, so Post attached heavier pontoons, salvaged from a much larger plane, to the Orion-Explorer. The pontoons extended so far forward that the aircraft became nose-heavy, and though it was manageable under normal conditions, there was a danger that it would dive out of control if the engine should fail. Post did not like the imbalance of the plane, but he thought he could handle it if Rogers sat well back toward the tail to counterbalance the overweight nose.

The two friends took off for Point Barrow on August 15, 1935. Bad weather threatened, but with the pontoons Post could, if he had to, land on any one of hundreds of small lakes. Less than 20 miles from their destination he decided to set down on a lake to ask directions; he was running short of fuel and did not want to risk getting lost. Eskimos gave him the information he sought, but meanwhile the plane's engine cooled. On takeoff, when the plane was about 50 feet above the lake, the engine stalled and the nose-heavy Orion-Explorer plunged into the water, snuffing out the lives of Wiley Post and Will Rogers.

America mourned its double loss. The crash itself was considered a severe setback to the increasing popularity of civilian air travel. Yet

In a picture that soon was published around the world, Will Rogers (left) chats with Wiley Post in Seattle just before the two friends left for Alaska in August 1935. "It's a 50-50 job," Rogers joked. "Wiley does the flying and I do the talking."

progress in aviation continued on many fronts. Pan American Airways began to carry mail and then passengers across the Pacific in big, dependable flying boats. Scheduled airline service between London and Australia had begun in April 1935 and would begin between the United States and Europe in 1939. The coming of World War II and the massive impetus it gave to aviation technology would finally make long-distance flights almost commonplace. Even before the War, there were signs that advancing science had supplanted the intuitive skill and heroic daring of the early airborne adventurers.

In 1938, a wealthy young film producer and aviator named Howard Hughes, with a crew of four, executed a meticulously planned world flight that obliterated Wiley Post's record. The Hughes plane was so well equipped and the ground arrangements so thorough that, although the flight evoked a rousing public response, the pilot refused to give himself much credit. Any airline pilot in a modern airplane, Hughes said, could have done the same thing. Pressed to compare his achievement with Post's, he chose his words with care: "Wiley Post's flight remains the most remarkable flight in history. It can never be duplicated." Then, echoing the tribute that had been paid to Lindbergh and to others among the pathfinders, Hughes added, "He did it alone!"

Post takes off from Seattle (below) to test the pontoons that were installed at the last minute on his Orion-Explorer. A few days later the plane lay on its back (bottom) after crashing in a lake near Point Barrow, Alaska, killing Post and Rogers.

Faultless dash that capped an era

When Howard Hughes's Lockheed 14 landed at New York's Floyd Bennett Field on July 14, 1938, the era of the pathfinders came to a definable close. The millionaire aviator and his crew of four had flown around the world in three days 19 hours 17 minutes, halving Wiley Post's five-year-old record. More significantly, their flight had gone precisely as planned: They made no unscheduled stops and never deviated more than a few miles from their prescribed course, demonstrating that long-distance flights were no longer the domain of daredevils.

Hughes had been planning the flight for three years, buying and discarding two other aircraft before he settled on the new Lockheed, a 12-passenger plane with twin 1,100-horsepower Wright Cyclone engines. His engineers and mechanics replaced the passenger seats with extra gas tanks and added so many sophisticated radios and navigational aids that the press called the plane "the Flying Laboratory." A chain of radio-equipped ships and ground stations was set up along the entire route to stay in constant touch with the plane. Critical spare parts were stored at the six scheduled stops as well as emergency sites in-between.

Hughes and his crew reached Paris in less than half the time it had taken Lindbergh in 1927, and from there it was on to Moscow, Omsk and Yakutsk in Siberia, where an improperly charted mountain range presented the trip's only unforeseen hazard. Once over the mountains, the flight proceeded without drama to Alaska, Minneapolis and back to New York. There a clamorous mob surged toward the plane—and discomfited the reticent Hughes more than anything that had happened in the air.

Howard Hughes (left) at 32 already held the world speed record of 352 mph and the record for the fastest transcontinental flight when he tackled his most ambitious aviation venture, investing a reported $300,000 in a flight around the world. Below, Hughes's Lockheed 14 taxis to a stop at the end of its record-setting trip.

Police at Floyd Bennett Field move in to contain a crowd that has broken through a restraining fence while awaiting Hughes's arrival. Frequent newspaper and radio reports of the flight's progress attracted a throng of 25,000 to the homecoming.

A cordon of police surrounds Hughes's plane in front of the airfield administration building, and city dignitaries assemble by the cabin door t

greet the airmen as they emerge after 91 hours of almost constant flying. The round-the-world speed record that they set stood for nine years.

Last to leave the plane, the six-foot-four-inch Hughes shakes the hand of New York's five-foot-two-inch mayor, Fiorello La Guardia.

Appendix

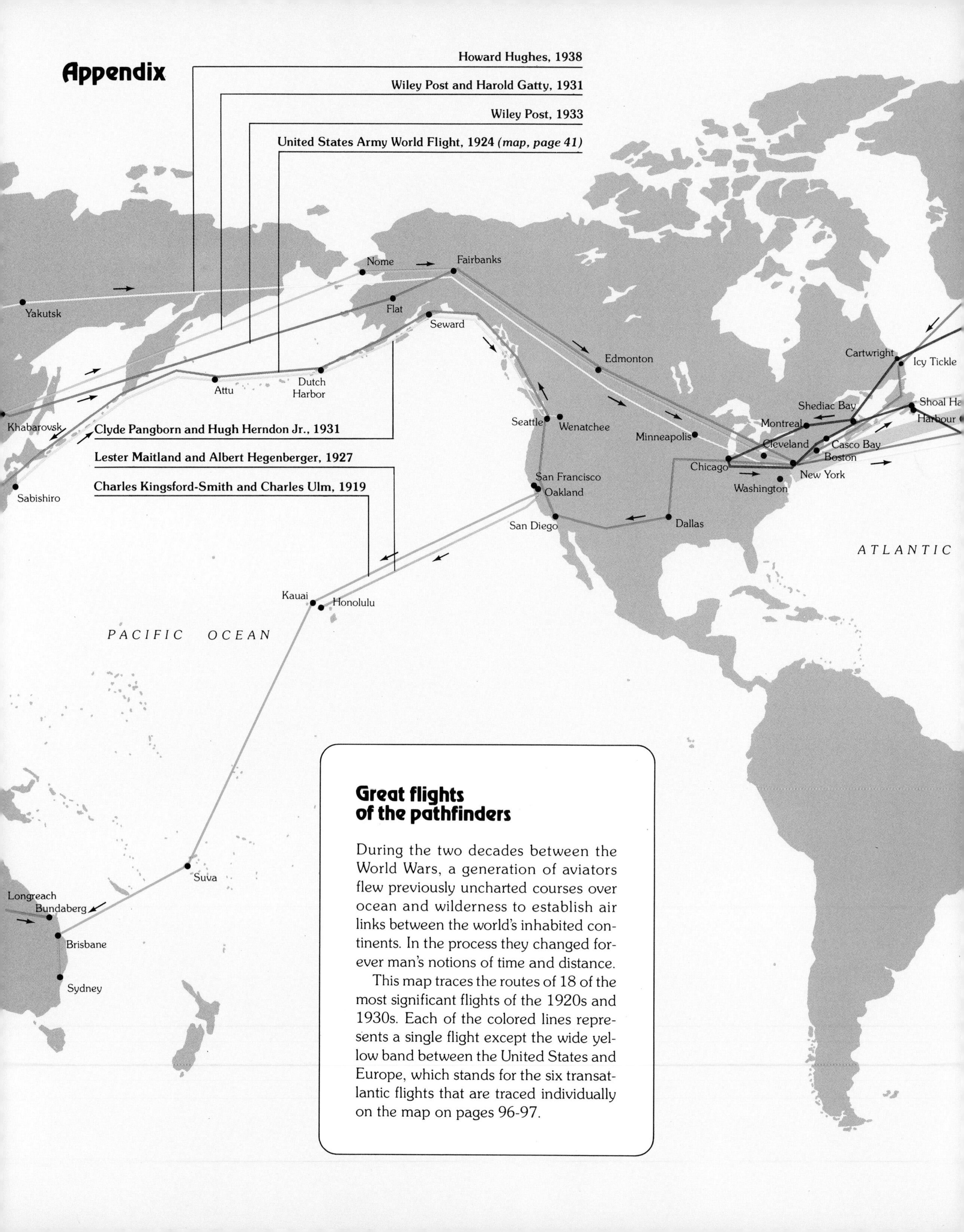

Great flights of the pathfinders

During the two decades between the World Wars, a generation of aviators flew previously uncharted courses over ocean and wilderness to establish air links between the world's inhabited continents. In the process they changed forever man's notions of time and distance.

This map traces the routes of 18 of the most significant flights of the 1920s and 1930s. Each of the colored lines represents a single flight except the wide yellow band between the United States and Europe, which stands for the six transatlantic flights that are traced individually on the map on pages 96-97.

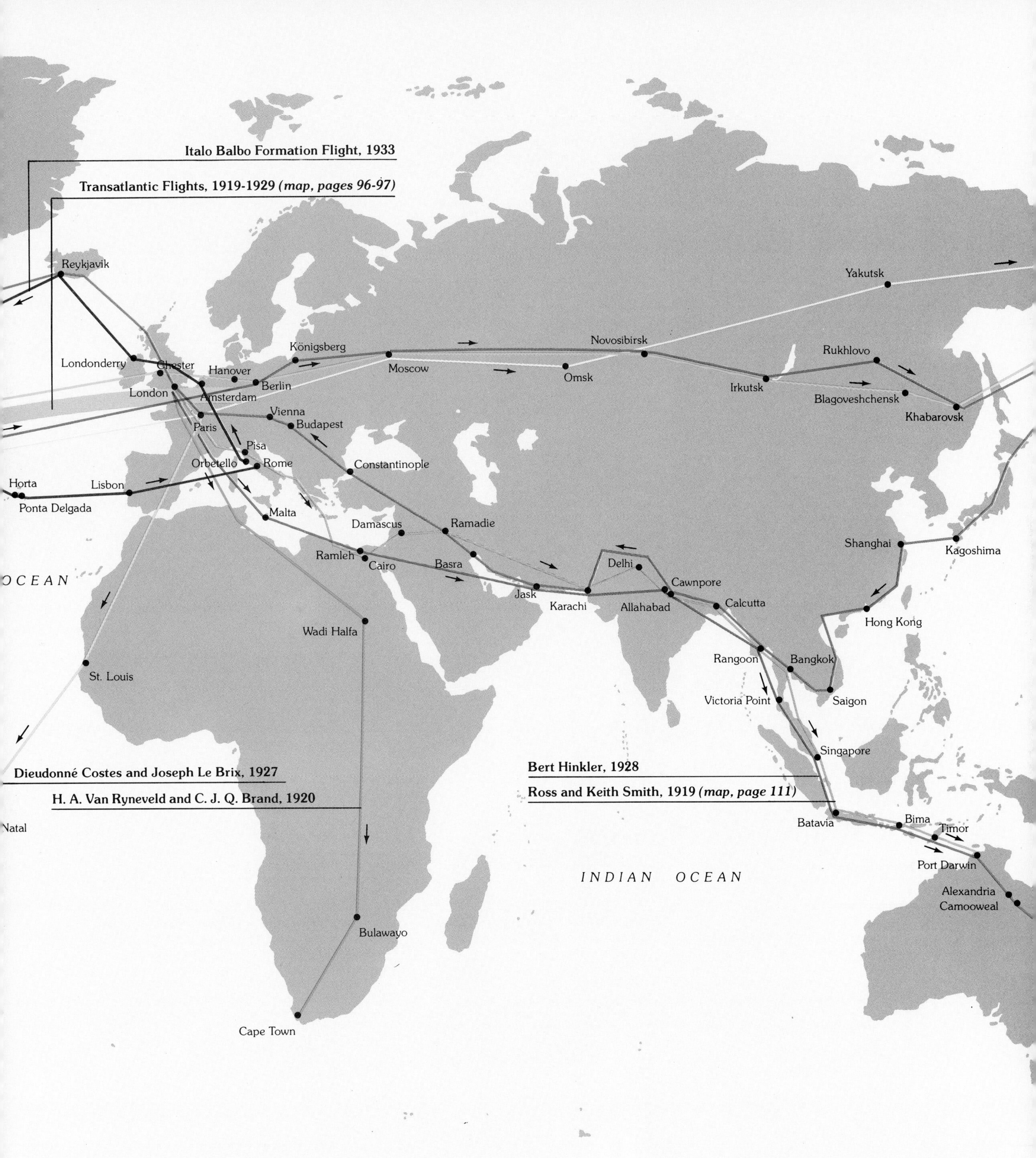

Italo Balbo Formation Flight, 1933
Transatlantic Flights, 1919-1929 (map, pages 96-97)
Reykjavik
Londonderry
Chester
London
Hanover
Amsterdam
Berlin
Königsberg
Moscow
Omsk
Novosibirsk
Yakutsk
Irkutsk
Rukhlovo
Blagoveshchensk
Khabarovsk
Paris
Vienna
Budapest
Pisa
Orbetello
Rome
Constantinople
Horta
Ponta Delgada
Lisbon
Malta
Damascus
Ramadie
Ramleh
Cairo
Basra
Jask
Karachi
Delhi
Cawnpore
Allahabad
Calcutta
Shanghai
Kagoshima
Hong Kong
Rangoon
Bangkok
Victoria Point
Saigon
Singapore
Batavia
Bima
Timor
Port Darwin
Alexandria
Camooweal
OCEAN
Wadi Halfa
St. Louis
Bulawayo
Cape Town
Natal
INDIAN OCEAN
Dieudonné Costes and Joseph Le Brix, 1927
H. A. Van Ryneveld and C. J. Q. Brand, 1920
Bert Hinkler, 1928
Ross and Keith Smith, 1919 (map, page 111)

Acknowledgments

The index for this book was prepared by Gale Linck Partoyan. The editors wish to thank John Batchelor, artist *(pages 72-73, 74-75, 152-153),* Paul Lengellé, artist *(front endpaper and cover detail, regular edition)* and Frank J. and Clare M. Ford, cartographers *(pages 41, 96-97, 170-171).*

For their valuable help with the preparation of this volume, the editors wish to thank the following: In Canada: Topsail, Newfoundland—Robert Plaskin; Ottawa—Robert Bradford, Curator of Aviation and Space, National Museum of Science and Technology; Joy Williams, National Photography Collection, Public Archives. In Denmark: Copenhagen—Bjorn Ocsner, Royal Library. In France: Nogent-Sur-Marne—Roland Nungesser, Député Maire; Paris—Gérard Baschet, Éditions de l'*Illustration;* Maurice Bellonte; Daniel Bouvard; Hervé Cras, Director for Historical Studies, Musée de la Marine; Odile de Langsdorff; André Bénard, Odile Benoist, Elisabeth Caquot, Lucette Charpentier, Alain Degardin, Gilbert Deloizy, Général Paul Dompnier, Deputy Director, Yvan Kayser, Général Pierre Lissarague, Director, Stéphane Nicolaou, Colonel Jean-Baptiste Reveilhac, Curator, Musée de l'Air; Colonel Marcel Dugué McCarty, Consultant, Lieutenant Colonel Marc Neuville, Curator, Colonel Paul Willing, Curator, Musée de l'Armée; Fabine Salates; Hélène Verlet, Curator, Bibliothèque Historique de la Ville de Paris; Vincennes—Général Charles Christienne, Director, S.H.A.A. In Great Britain: Cranwell—Jean King, Royal Air Force College; Dorset—Terry Treadwell; Hampshire—A. S. Lumsden; Hendon—Anthony Harold, Ray Lee, Richard Simpson, Royal Air Force Museum; Kent—David Beaty; London—J. B. Marriott, Keeper of the Royal Philatelic Collection; A. W. L. Nayler, Royal Aeronautical Society; Steve Piercey, Flight International; Martin Andrewartha, John Bagley, Science Museum; Ian Stokkes, Vickers; Marjorie Willis, BBC Hulton Picture Library; Manchester—W. T. Charnock; N. Edwards; Norwich—Morag Barton, Weybridge Museum; Chaz Bowyer; Anthony Hancocks, Alan Jeffcoate, Vintage Aircraft Flying Association. In Italy: Milan—Maurizio Pagliano, Rizzoli; Rome—Paolo Balbo; Countess Maria Fede Caproni, Museo Aeronautico Caproni di Taliedo; Colonel Gennaro Adamo, Captain Giovanni Angelini, Ufficio Storico, Captain Giancarlo Fortuna, Stato Maggiore Aeronautica. In South Africa: Johannesburg—Arthur Blake; Major Peter McGregor, Director, South African Air Force Museum. In the United States: Arizona—Harry Weisberger, Sperry Flight Systems; California—Robert C. Ferguson; David Hatfield, Northrop University; Ed Lund; H. Edwin Morrow; William Wagner, Ryan Aeronautical Library. In Connecticut—John A. Cox, Pratt & Whitney Aircraft Group; H. H. Lippincott, Archivist, UTC Archives; Judith A. Schiff, Yale University. In Washington, D.C.—Agnes Hoover, Naval Historical Center; Leroy Bellamy, Mary Ison, Jerry Kearns, Bernard Reilly, Prints and Photographs Division, Library of Congress; Richard Thurm, Paul White, Audio-Visual Division, National Archives and Records Service; Richard D. Crawford, Tom Crouch, Phil Edwards, General Benjamin S. Kelsey, USAF Ret., Robert B. Meyer Jr., Dominick A. Pisano, Mimi Schaef, Pete Suthard, C. Glenn Sweeting, National Air and Space Museum, Smithsonian Institution; Don Fisher, Smithsonian Press; Andrew Parker; General and Mrs. Leigh Wade, USAF Ret. In Massachusetts—The Institute Archives and Special Collections, Massachusetts Institute of Technology; The Massachusetts Institute of Technology Historical Collections. In Missouri—Charles Nelson, Midcoast Aviation; Susan A. Riggs, Judith Schwartz, Missouri Historical Society. In New Jersey—Ted Lustig, Sun Chemical Corporation. In New York—William Kaiser, Curator, Cradle of Aviation Museum; New York City—Mrs. Hugh Herndon Sr.; Elizabeth Mason, Butler Library, Columbia University; Alexander R. Ogston, The Wings Club Inc.; Arthur E. Sauvigné, Sperry Corporation. In Ohio—Dr. Stanley R. Mohler, Wright State University; Katherine E. Cassity, Royal Frey, Curator, Charles G. Worman, Air Force Museum, Wright-Patterson Air Force Base. In Oklahoma—Paul English, Curator, Oklahoma Historical Society; John Hill, Curator, Oklahoma Territorial Museum; Bob Lansdown, Curator, Woolaroc Museum; Mary Jane Norfleet, Phillips Petroleum Company. In Pennsylvania—Ted Hake, Hake's Americana and Collectibles. In Washington—Hu Blonk, Wenatchee Daily World; Carl M. Cleveland; Beverly South, William E. Steward, Director, North Central Washington Museum; Karl Stoffel. In Virginia—Dana Bell, USAF Still Photo Depository; Richard Stephenson, Geography and Maps Division, Library of Congress. In West Germany: Berlin—Dr. Roland Klemig, Heidi Klein, Bildarchiv Preussischer Kulturbesitz; Wilma Frielingsdorf; Axel Schulz, Eva Bong, Ullstein Bilderdienst; Deisenhofen—Josef Pöllitsch; Koblenz—Dr. Matthias Haupt, Meinhard Nilges, Bundesarchiv.

The editors also wish to thank Enid Farmer, Cambridge, Massachusetts; Janet Zich, Half Moon Bay, California; Diane Asselin, Los Angeles; Carol Barnard, Seattle.

Particularly useful sources of information and quotations were *Our Transatlantic Flight* by Sir John Alcock and Sir Arthur Whitten Brown, William Kimber, Ltd., London, 1969; *The Spirit of St. Louis* by Charles A. Lindbergh, Charles Scribner's Sons, 1953; *Around the World in Eight Days* by Wiley Post and Harold Gatty, Rand McNally & Company, 1931.

Bibliography

Books

Alcock, Sir John, and Sir Arthur Whitten Brown, *Our Transatlantic Flight.* London: William Kimber, 1969.

Allen, Richard Sanders, *Revolution in the Sky, Those Fabulous Lockheeds, the Pilots Who Flew Them.* The Stephen Greene Press, 1964.

Balbo, Italo, *My Air Armada.* London: Hurst & Blackett, 1934.

Barlett, Donald L., and James B. Steele, *Empire, The Life, Legend and Madness of Howard Hughes.* W. W. Norton, 1979.

Brown, Arthur Whitten, *Flying the Atlantic in Sixteen Hours.* Stoke, 1920.

Byrd, Richard Evelyn, *Skyward.* G. P. Putnam's, 1928.

Chamberlin, Clarence D., *Record Flights.* Dorrance Company, 1928.

Cleveland, Carl M., *"Upside-Down" Pangborn, King of the Barnstormers.* Aviation Book Company, 1978.

Crouch, Tom D., *Charles Lindbergh, An American Life.* National Air and Space Museum, 1977.

Davis, Pedr, *Charles Kingsford Smith, the World's Greatest Aviator.* Sydney: Paul Hamlyn Pty Limited, 1978.

Doolittle, James H., *Early Experiments in Instrument Flying.* Smithsonian Institution, 1962.

Ellis, F. H. and E. M., *Atlantic Air Conquest.* London: William Kimber, 1963.

Eustis, Nelson, *The Greatest Air Race.* Sydney: Rigby Limited, 1969.

Fleischer, Suri, and Arleen Keylin, eds., *Flight, as Reported by The New York Times.* Arno Press, 1977.

Fokker, Anthony H. G., and Bruce Gould, *Flying Dutchman, the Life of Anthony Fokker.* Arno Press, 1972.

Gibbs-Smith, Charles Harvard, *Aviation.* London: Science Museum, 1970.

Guggenheim, *Equipment Used in Experiments to Solve the Problem of Fog Flying.* The Daniel Guggenheim Fund for the Promotion of Aeronautics, 1930.

Hallion, Richard P., *Legacy of Flight, the Guggenheim Contribution to American Aviation.* University of Washington Press, 1977.

Hamlen, Joseph R., *Flight Fever.* Doubleday, 1971.

Hawker, H. G., and K. Mackenzie Grieve, *Our Atlantic Attempt.* London: Methuen & Co., 1919.

Jablonski, Edward, *Atlantic Fever.* Macmillan, 1972.

Jane, Fred T., *Jane's All the World's Aircraft, 1938.* Compiled and ed. by C. G. Grey and Leonard Bridgman. Arco Publishing Company, 1938.

Kingsford-Smith, C. E., and C. T. P. Ulm, *The Flight of the Southern Cross.* Robert M.

McBride & Company, 1929.
Lindbergh, Charles A.:
Autobiography of Values. Harcourt Brace Jovanovich, 1977.
The Spirit of St. Louis. Charles Scribner's, 1953.
We. G. P. Putnam's, 1927.
McDonough, Kenneth, *Atlantic Wings.* Hemel Hempstead: Model Aeronautical Press, 1966.
Mohler, Stanley R., and Bobby H. Johnson, *Wiley Post, His Winnie Mae, and the World's First Pressure Suit.* Smithsonian Institution Press, 1971.
Mosley, Leonard, *Lindbergh.* Dell, 1976.
Post, Wiley, and Harold Gatty, *Around the World in Eight Days.* Rand McNally, 1931.
Roseberry, C. R., *The Challenging Skies.* Doubleday, 1966.
Ross, Walter S., *The Last Hero: Charles A. Lindbergh.* Harper & Row, 1976.
Schlaifer, Robert, and S. D. Heron, *Development of Aircraft Engines* and *Development of Aviation Fuels.* Graduate School of Business Administration, Harvard University, 1950.
Sikorsky, Igor, *The Story of the Winged-S.* Dodd, Mead, 1939.
Stark, Howard C., *Blind or Instrument Flying Instruction Book.* Howard C. Stark, 1932.
Thomas, Lowell, *The First World Flight.* London: Hutchinson & Co., 1926.
Turner, P. St. John, *The Vickers Vimy.* London: Patrick Stephens, 1969.
Villard, Henry Serrano, *Contact! The Story of the Early Birds.* Bonanza Books, 1968.
Wagner, William, *Ryan, the Aviator.* McGraw-Hill, 1971.
Wright, Monte Duane, *Most Probable Position, A History of Aerial Navigation to 1941.* The University Press of Kansas, 1972.

Periodicals
"*Air-Cooled Engines.*" Great Britain Air Ministry, *Air Publication 800,* November 1920.
Diamond, Emanuel, "Atlantic Fliers Await Word 'Go.' " *The New York Times,* August 22, 1926.
"Flight of Ye Good U.S. Ship NC-4." *Aviation Quarterly,* Vol. 2, Nos. 1-4, 1976.
"Fonck Plane 'Prettied Up' As Weather Halts Tests." *Syracuse Herald* (Syracuse, N.Y.), August 15, 1926.
"Gatty's Navigation Instrument." *Aero Digest,* January 1932.
McLaughlin, George, "The All-Metal Sikorsky." *Aero Digest,* June 1926.
"Nine Days Around the World." *Aviation Magazine,* August 1931.

Picture credits

The sources for the illustrations that appear in this book are listed below. Credits for the illustrations from left to right are separated by semicolons; from top to bottom they are separated by dashes.
Front endpaper (and cover detail, regular edition): Painting by Paul Lengellé.
7: Jean-Loup Charmet, courtesy Musée de l'Air, Paris; inset, Mary Evans Picture Library, London. 8: Dmitri Kessel, courtesy Musée de l'Air, Paris. 9: Painting by Achille Beltrame, courtesy *Domenica del Corriere,* Milan; inset, S.A.F.A.R.A. Collection, Bibliothèque Nationale, Paris. 10: Painting by Achille Beltrame, courtesy *Domenica del Corriere,* Milan; inset, Musée de l'Air, Paris. 11: Dmitri Kessel, courtesy Musée de l'Air, Paris; inset, Radio Times Hulton Picture Library, London. 12: Photo Bibliothèque Nationale, Paris. 13: Jean-Loup Charmet, courtesy Private Collection, Paris.
14: Derek Bayes, courtesy British Airports Authority. 16: Derek Bayes, courtesy Mary Evans Picture Library, London. 17: Derek Bayes, courtesy Science Museum, London. 18, 19: National Air and Space Museum, Smithsonian Institution. 20: Henry Beville, from *The Flight of Alcock and Brown* by Graham Wallace, published by Putnam, London. 21: Canadian Armed Forces, courtesy Newfoundland Museum, St. John's. 23: Charles Phillips, courtesy National Air and Space Museum, Smithsonian Institution. 24: Newfoundland Museum, St. John's. 25: From *Newfoundland Airmails 1919-1939,* courtesy Harmers of London Ltd.—reproduced by the gracious permission of Her Majesty Queen Elizabeth II. 27: Newfoundland Museum, St. John's—Royal Air Force College, Cranwell. 28: Vickers, London—Royal Air Force College, Cranwell. 29: Newfoundland Museum, St. John's—Royal Air Force College, Cranwell.
31: Vickers, London. 32: The Mansell Collection, London. 34: Derek Bayes, courtesy Royal Air Force Museum, Hendon—Derek Bayes, courtesy Manchester Airport Authority. 35: Derek Bayes, courtesy Royal Air Force Museum, Hendon.
38: Éditions de l'*Illustration,* Paris. 39: Vickers, London. 40, 41: National Air and Space Museum, Smithsonian Institution; inset, map by Frank J. and Clare M. Ford. 42: Paulus Leeser, courtesy U.S. Air Force Museum—U.S. Air Force Photo Depository. 43: U.S. Air Force Photo Depository. 44: Paulus Leeser, courtesy Lowell Thomas Collection—Paulus Leeser, courtesy U.S. Air Force Museum. 45: Paulus Leeser, courtesy Lowell Thomas Collection. 46: U.S. Air Force Photo Depository. 47: Paulus Leeser, courtesy U.S. Air Force Museum—Paulus Leeser, courtesy Lowell Thomas Collection. 48: UPI. 49: Paulus Leeser, courtesy U.S. Air Force Museum—U.S. Air Force Photo Depository. 50: Dmitri Kessel, courtesy Musée de l'Air, Paris. 52: United Technologies Archives. 53: UPI. 54: National Air and Space Museum, Smithsonian Institution—Hatfield History of Aeronautics, Northrop University. 57: The Bettmann Archive—Charles Lindbergh Papers, Yale University Library. 58: Charles Lindbergh Papers, Yale University Library, except bottom right, Minnesota Historical Society. 59: Charles Lindbergh Papers, Yale University Library. 60, 62: Missouri Historical Society. 65, 68: UPI. 71: Ryan Aeronautical Library. 72-75: Drawings by John Batchelor. 76: UPI. 78: Photographer unknown, courtesy Musée de l'Air, Paris. 81: Musée de l'Air, Paris. 82: Musée de l'Air, Paris, except bottom left, Missouri Historical Society. 85: The Bettmann Archive. 87: National Museum of Science and Technology, Ottawa—Bundesarchiv, Koblenz. 88: Cradle of Aviation Museum, Nassau County, N.Y. 89: Wide World. 90: Painting by Gerald Coulson—painting by William Reynolds.
91: Painting by Charles Hubbell, by permission of TRW Inc.; Charles Phillips, by permission of Atlantic Aviation Corporation, painting by Keith Ferris—painting by John T. McCoy, Aviation Americana Inc. 92: Missouri Historical Society.
94: Brown Brothers. 96, 97: Map by Frank J. and Clare M. Ford. 98, 99: The Bettmann Archive.
100: Naval Historical Center—Henry Groskinsky, courtesy Missouri Historical Society. 101: The Bettmann Archive. 102, 103: National Air and Space Museum, Smithsonian Institution.
104: *New York Daily News.* 105: Missouri Historical Society—Henry Groskinsky, courtesy Missouri Historical Society (3)—National Air and Space Museum, Smithsonian Institution. 106: Henry Groskinsky, courtesy Missouri Historical Society (3)—Henry Groskinsky, courtesy Hake's Americana & Collectibles, York, Pa. (4)—courtesy Bella Landauer Sheet Music Collection, National Air and Space Museum, Smithsonian Institution, photos by Fil Hunter, except center, Charles Phillips. 107: Henry Groskinsky, courtesy Missouri Historical Society. 108: Kevin Berry, courtesy Sydney Airport. 111: Qantas, Sydney.
113: Vickers, London. 115: Selwyn Tait, courtesy South African Air Force Museum, Pretoria.
116: Radio Times Hulton Picture Library, London. 117: Port Collection, Queensland Museum, Brisbane. 118: The Mansell Collection, London. 120, 121: Hawaii State Archives. 122: Associated Newspapers, London. 124: Top, The Bettmann Archive. 128, 129: Photoworld. 130-137: Stato Maggiore Aeronautica, Rome. 138: The Bettmann Archive. 141: Lufthansa, Cologne. 142, 143: A. Y. Owen, courtesy Oklahoma Historical Society. 145: Courtesy Mrs. Hugh Herndon Sr. (2)—Tom Tracy, courtesy North Central Washington Museum—National Air and Space Museum, Smithsonian Institution (2). 147: Lockheed Corporation. 148: U.S. Air Force Museum.
149: Sperry Flight Systems Division, Sperry Corp., except center, Kollsman Instrument Company. 152, 153: Drawing by John Batchelor. 154: National Archives. 306NT-336-JJ-7. 156: Lockheed Corporation. 158: A. Y. Owen, courtesy Oklahoma Historical Society. 160: Wide World.
161: UPI—National Air and Space Museum, Smithsonian Institution. 162, 163: Culver Pictures—The Bettmann Archive. 164-167: *New York Daily News.* 168, 169: Wide World. 170, 171: Map by Frank J. and Clare M. Ford.

Index

Numerals in italics indicate an illustration of the subject mentioned.

A

Acosta, Bert, 97
Aeronautical engineers, 54, 65
Aeroplane magazine, 16
Airmail pilots (U.S.), 52, 55, 61-63, *62,* 119
Alaska: 160-*161;* and round-the-world flights, 150, 157
Alcock, John "Jack," 16, 21, 24-26, 27, 30-33, 36-39, 51, 52, 92, *map* 96-97, 110; death of, 39; Heathrow monument to, *14;* landing of, *38;* prize to, *39;* quoted, 18, 20, 26, 30, 33, 37, 38; takeoff of, *31, 32;* waistcoat of, *35*
Alps mountains: Balbo and, *132-133;* crash over, *9*
America (Fokker Trimotor), 64, 72, *73,* 86, 97
American Legion (Keystone Pathfinder), 64, 67, *72;* crash of, 67-*68,* 72
Amundsen, Roald, *124*
Anderson, Keith, 121-123
Arnold, Lieutenant Leslie P., *42;* quoted, 44
Atlantic (Sopwith plane), 18-20, 22, 24, *25-26*
Atlantic Ocean, *map* 96-97, *map* 170-171; Balbo and, *130-137;* first east-west flight across, *87;* NC's and, *23;* shortest distance across, 16-18, 52. *See also* Orteig Prize; Prizes
Australia, 109; flight across Pacific to, 118, 119, 121, 122-123, 125, 127, 129; Hinkler's flight to, 116-117; Kingsford-Smith and, 119-123, 125-127, 129; MacRobertson race to, 159, 160; perimeter flight around, 122; Smiths' flight to, 110-114
Australian Flying Corps, 110, 111
Avro planes, 16, 116; Avian biplane, 116-*117,* 129; Baby, 116
Azores, 23, *map* 96-97

B

Bailey, Mary, 139
Balbo, Italo, *130-137, map* 170-171
Balchen, Bert, 97, 125
Bangkok, Thailand, 113, *map* 170-171
Barnstorming, 141, 142; Australian pilots and, 119, 120, 123; U.S. pilots and, 52, 55, 56, *60,* 61, 77
Batavia (Djakarta), Indonesia, 114, 117, *map* 170-171
Beaumont, André, *11*
Bellanca, Giuseppe, 54, 64, 65, 67, 86. *See also Columbia* (Wright-Bellanca monoplane); Wright-Bellanca monoplanes
Bennett, Floyd, *124,* 125; and Orteig Prize, 64, 68
Bennett, Jim, 111, 112, 113, 114
Benoist, Jean, 110-112, 114
Berlin, Germany, *map* 96-97, 146, 155, 156; Brandenburg Gate, *141*
Bima, 114, *map* 170-171
Biplanes, 72; Avro Avian, 116-117, 129; Bellanca, 54; Breguet, 139; Bristol Tourer, 122; and Dole Race, 123; Douglas World Cruisers, *40-49;* Dove (Sopwith), 109, 114, 116; Farman, 16; Lincoln Standard, 56, *59;* monoplanes versus, 54; *Sunrise,* 19. *See also Atlantic* (Sopwith plane); *L' Oiseau Blanc;* Sikorsky S-35 (biplane); Vimy (Vickers bomber)
Bird of Paradise (Fokker Trimotor), 118-119
Blackburn bomber, 110
Blériot, Louis, 6, *7;* monoplanes, *6, 7, 9*
Blind flying, 144, *148-149,* 155, 157. *See also* Mechanical Mike (autopilot system)
Bly, Nellie, 140
Blythe, Dick, 84, 86, 88; quoted, 84
Boardman, Russell, 139
Bombers: Blackburn, 110. *See also American Legion* (Keystone Pathfinder); Handley Page, Ltd.; Vickers, Ltd.
Boston (Douglas World Cruiser), 40, 43, *47*
Boston, Mass., 81, *map* 170-171
Bowlus, Hawley, 70
Brakes, absence of, 53, 114
Brand, Flight Lieutenant C. J. Q., 115, *map* 170-171; flag to, *115;* medicine chest of, *115*
Brazil, 139
Breguet biplane, 139
Bremen (Junkers monoplane), *87*
Brisbane, Australia, *117,* 125, 129, *map* 170-171
Bristol planes, 16; Tourer, 122
British Columbia, 140
British War Mission (New York), 22
Brooklands, England, aviation school at, 16, 17, 18, 22, 25
Brooks Field (San Antonio, Texas), 60-61
Brower, Charlie, 160
Brown, Arthur Whitten "Teddie," 17-21, *20,* 24-25, 27, 30-33, 36, 51, 52, 92, *map* 96-97, 110; Heathrow monument to, *14;* jacket of, *35;* landing of, *38;* mascot of, *35;* prize to, *39;* quoted, 32, 33, 36-37, 38; sextant of, *34;* takeoff of, *32*
Bruno, Harry, 84, 88; quoted, 89
Brussels, Belgium, 96
Byrd, Commander Richard E., 64, 67, 68, 72, 80, 81, 86, 88, *map* 96-97, 123; crash of, 80; polar flights of, 64, *124*-125; quoted, 124. *See also America* (Fokker Trimotor)

C

Canada, and round-the-world flights, 150-151, 157
Canuck (Canadian Jenny), 141, 142
Cape Town, South Africa, 115, *map* 170-171
Caudron G.4, 109, 110, 112
Chamberlin, Clarence, 54, 65, 69, 80, 81, 86, *88, map* 96-97; quoted, 54
Charleville, Queensland, Australia, postcard from, *111*
Chavez, Georges, *9;* quoted, 9
Cherbourg, France, 95, 97, 140
Chicago (Douglas World Cruiser), 40, 43
Chicago, Ill., *map* 170-171; airmail and, 61, 63; World's Fair (1933), 130, *134-135*
Churchill, Winston, *39;* quoted, 39
Circuit de l'Est (French race), *8*
Circuit of Britain and Europe races, 11
Clavier, Charles, 53-54
Cleveland, Ohio, 39, 151, *map* 170-171
Clifden, Ireland, 38, *map* 96-97
Coli, François, 67, 71, 78-80, *81;* disappearance of, 72, 81, *82,* 85-86, 96; quoted, 80. *See also L'Oiseau Blanc*
Collier Trophy, 55
Collyer, Charles, 140
Columbia (Wright-Bellanca monoplane), 65, 67, 69, *73,* 80, 86, 97
Columbia Aircraft Company, 65
Consolidated NY-2 test plane, *148-149*
Coolidge, Calvin, *49, 101;* quoted, 49
Costes, Dieudonné, 139, *map* 170-171
Coulson, Gerald D., painting by, *90*
Cranwell, England, 139
Crashes and accidents, 56, 61, 62-63, 109-110, 122, 140; *American Legion, 68;* and Australia, 114, 121; Balbo's planes, 130; barnstormers and, 60; Chavez', *9;* and Dole Race, 123; and Douglas World Cruisers, *47;* Mattern's, *155;* and Northcliffe's prize, *18-19, 24, 25,* 39; and Orteig Prize, *53,* 54, 55, 63, 67-69, 72, 78, 79, 80, 88; of Post and Rogers, 160-*161;* and round-the-world flights, 155, 157, 159; *Silver Queen II,* 115; Védrines', *12*
Curtin, Lieutenant Lawrence W., 53-54
Curtiss Field (Long Island, N.Y.), 53; Lindbergh at, 84, *85,* 86, *89*
Curtiss planes: Hawk fighter, 71; NCs (flying boats), *23;* pusher, 141. *See also* Jennies
Cyclone engines (Wright), 143, 162

D

Daily Mail, The (London newspaper), 16, 25, 26
Darwin, Australia, 114, 117, *map* 170-171
Davis, Lieutenant Commander Noel, 64, 67; death of, *68,* 72, 80; quoted, 67. *See also American Legion* (Keystone Pathfinder)
De Havilland planes, 52, 55; D.H.6s, 119; D.H.9, 115
Diggers' Aviation Ltd. (Australia), 120-121
Dirigible, *Norge,* 124. *See also Graf Zeppelin* (German dirigible)
Distinguished Flying Cross (U.S.), *100,* 130
Dole Race, 123
Dooley, Agnes, 21, 22, 32, 36; quoted, 20
Doolittle, Lieutenant James, 51; blind flight of, *148-149*
Douglas, Donald, 40
Douglas Aircraft, World Cruisers, *40-49, map* 41
Doumergue, Gaston, 80
Dove (Sopwith biplane), 109, 114, 116
Ducrocq, Maurice, 16
Dunlap, Vera May, Flying Circus broadside, *60*

E

Earhart, Amelia, 139
Engines: air-cooled radial, 52-53, 54, 55; Isotta-Fraschini Asso, 130; Jupiter, 52, 72; Liberty, 23; Lorraine-Dietrich, 79. *See also* Cyclone engines (Wright); Rolls-Royce engines; Wasp engine (Pratt & Whitney); Whirlwind engines (Wright)
English Channel, 6, *7,* 51, 64, 81
Evergreen Tree (Sioux chief), *135*

F

F. VII (Fokker plane), 70
Fairchild monoplane, 140
Farman biplane, 16
Fédération Aéronautique Internationale, 110, 142
Ferris, Keith, painting by, *91*
Fiji Islands. *See* Suva, Fiji Islands
Fitzmaurice, James, *87*
Floyd Bennett Field (Long Island, N.Y.), 155, 157; Hughes and, *162-169*
Flying boats: Balbo's, *130-137;* Curtiss (NCs), *23;* PN-9, 118
"Flying Laboratory, The" (Hughes's plane), 162
Foch, Marshal Ferdinand, 96
Fokker, Anthony, 68, 88
Fokker company: F.VII, 70; Lindbergh and, 63-64; *St. Raphael,* 64. *See also* Trimotors (Fokker planes)
Fonck, René: and Lindbergh, 86, 88; and Orteig Prize, 52, *53*-54, 55, 63, 72, 78; quoted, 54. *See also* Sikorsky S-35 (biplane)
Ford Tri-motor, *Floyd Bennett, 124*
France, competition with U.S., 51-52
French flying circus, tour of U.S., 78
Friedrichshafen, Germany, 140

G
Garros, Roland, *13*
Gatty, Harold, 144, 146, 150-151, 154, 157, 159, *map* 170-171; instrument invented by, *152-153;* quoted, 146
George V, King (England), 39; letter to, *25*
Gibbons, Floyd, *154*
Gilbert, Eugène, *10*
Goodrich, B. F., Rubber Company, 159
Graf Zeppelin (German dirigible), 140, *141,* 151
Greenly Island (Labrador), *87, map* 96-97
Grieve, Lieutenant Commander Kenneth Mackenzie "Mac," *21-22,* 25-26
Guggenheim, Daniel, Fund for the Promotion of Aeronautics, 148
Gurney, Bud, *59*

H
Hall, Donald, 65, 66-67, 70, 71; quoted, 66, 81
Hall, F. C., 142-143, 144, 151
Hancock, Captain G. Allan, 123, 125, 129
Handley Page, Ltd.: bomber, 110; V/1500 bomber, 19-20, 26, 30, 39
Harbour Grace (Newfoundland), 20, 21, 26, 30, *map* 170-171
Harding, Lieutenant John, Jr., 42, *43*
Harvey, Sergeant Alva L., *42*
Hawaii, 139; Dole Race, 123; Kingsford-Smith and, 125-126; U.S. flights to, 118-119, 123; Wheeler Field, 119, *120-121,* 125-126
Hawker, Harry, 18-22, *21,* 24-26, 32; quoted, 20, 22, 25
Heath, Mary, 139
Hegenberger, Lieutenant Albert, 118-119, *120-121,* 123, *map* 170-171
Herndon, Hugh, Jr., 144-*145*
Herrick, Ambassador Myron T., 96
High-altitude flying, 159-161
Hinkler, Lieutenant Bert, 109, 110, 114; flight to Australia, *116*-117, *118,* 129, *map* 170-171
Hinton, Lieutenant Walter, quoted, 23
Hubbell, Charles H., painting by, *91*
Huenefeld, Guenther von, *87, map* 96-97
Hughes, Howard, *162-169, map* 170-171; quoted, 161
Hutchinson, Rear Admiral Benjamin, *100*

I
Icy Tickle (Labrador), *48, map* 170-171
India, 109, 111, 112, 139; Douglas World Cruisers and, *45*
Indonesia, 109; Surabaya, *113,* 114. *See also* Batavia (Djakarta), Indonesia
Iraq, 112, 116
Ireland, 17, 19, 22, 30, 36, 37, *38,* 81, 95, 96
Irish Sea, Wood's crash into, *18-19*
Isotta-Fraschini Asso engines, 130

J
Japan, 140; *Asahi Shimbun* prize, 144-*145;* Douglas World Cruisers and, *44*
Jennies (Curtiss JN-4 trainers), 52, 56, 60; Canuck (Canadian version), 141, 142
Junkers monoplane *(Bremen), 87*
Jupiter engines, 52, 72

K
Kelly, Mayor Edward J., *135*
Kelsey, Lieutenant Ben, 148
Kerr, Admiral Mark, 19, 20, 21, 26, 30, 39
Keystone Aircraft Corporation, 64
Kimball, James H., 81, 86, 92, 146
Kingsford-Smith, Charles, *108,* 110, 119-123, *122,* 125-127, 129, 139, *map* 170-171; quoted, 125, 129
Kingsford-Smith, Maddocks Aeros Ltd., 119
Koehl, Hermann, *87*
Kollsman, Paul, instrument devised by, *149*

L
La Guardia, Fiorello, *168-169*
Lady Southern Cross (Lockheed Altair), 129
Lambert Field (St. Louis, Mo.), 61, 81
Lang, J. T., 122, 123
Langley Field, Va., 67; crash at, *68*
Le Blanc, Alfred, *8*
Le Bourget airfield (France), 80-81, *94,* 95-96
Le Brix, Joseph, 139, *map* 170-171
Lengellé, Paul, painting by, *front endpaper*
Levasseur, Pierre, 78; quoted, 80, 82. *See also L'Oiseau Blanc*
Levine, Charles A., 65, 67, 69, 97
Liberty engines, 23
Lincoln Standard biplane, 56, *59*
Lindbergh, Charles Augustus, 55-64, *57, 62,* 67, 69-*71, 85, 89,* 146, 161; gifts to, *106-107;* New York to Paris flight, *76,* 89-96, *map* 96-97, 118; Nungesser and, 78, 81; parents of, *57;* post-flight acclaim, *98-107;* quoted, 55, 56, 61, 62, 63, 66, 67, 69, 70-71, 81, 83, 88, 89, 92, 94, 95, 100; San Diego to New York flight, 81, 83-86, 88-89; search for plane, 63-67; youth of, *57-59. See also* Orteig Prize, Lindbergh and; *Spirit of St. Louis*
Lockheed planes: Altair, 129; Explorer, 160-*161;* Hughes's, *162-169;* Orion, 160-*161;* Vega, 143-144, 155. *See also Winnie Mae*
L'Oiseau Blanc (PL-8), 67, *73, 78-81;* disappearance of, 72, 81, 85-86
London, England, *map* 170-171; and Australia, 110-114, 129; Lindbergh in, 96-97; Savoy Hotel ceremony, *39; Times,* 39.
Lorraine-Dietrich engine, 79
Lowenstein-Wertheim, Princess Anne, 64-*65*
Lynch, H. J. "Cupid," 56
Lyon, Harry W., 125, 126, 129; quoted, 125

M
MacRobertson Race, 159, 160
Maddocks, Cyril, 119, *122*
Mahoney, Franklin, 65
Maitland, Lieutenant Lester, 118-119, *120-121,* 123, *map* 170-171; quoted, 118
Martin, Major Frederick L., *42, 44*
Martinsyde Company, 16, 19; *Raymor,* 19, 22, *24,* 26, 30
Mary (Danish tramp steamer), 25, 26
Mattern, James J., 155, *156*
Matthews, Captain George Campbell, 110-111, 114
Maughn, Lieutenant Russell A., 51-52
Mauretania (British passenger ship), 20
McCoy, John T., painting by, *91*
Mears, John Henry, 140
Mechanical Mike (autopilot system), 155, 156-157, 159
Memory, Fred, quoted, 25
Memphis, U.S.S. (U.S. ship), 96-97, *100*
Minchin, Fred, 64
Miss Veedol (Wright-Bellanca monoplane), 144-*145*
Mitchel Field (Long Island, N.Y.), 148
Mitchell, General William "Billy," 51
Moffett-Starkey Aero Circus, 119
Monoplanes, 72; versus biplanes, 54; Blériot, *6, 7, 9;* Fairchild, 140; Junkers *(Bremen), 87. See also America* (Fokker Trimotor); *Columbia* (Wright-Bellanca monoplane); *Spirit of St. Louis;* Wright-Bellanca monoplanes
Morgan, Captain C. W. Fairfax "Fax," 19, 20, *21, 24*-25; quoted, 22
Morrow, Ed, quoted, 69, 71, 83
Mount Pearl (Newfoundland), 20, 21, 25
Muller, Maxwell "Max," 17-18, 30; quoted, 18
Mussolini, Benito, 130, 131, *137*

N
National Advisory Committee for Aeronautics (NACA), 143
National Air Races, Men's Air Derby, 143-144
Navigation devices, 15, 17, 33, *34,* 36, 37, 118-119, 125-126, 144; Hughes and, 162; Lindbergh and, 70, 84, 86. *See also* Blind flying; Mechanical Mike (autopilot system)
Navigators: and Australian flights, 110, 111, 118-119, 122, 125; and Northcliffe's prize, 17-18, 19; and Orteig Prize, 63, 64, 66, 67, 70, 78-79; and round-the-world flight, 144, 146. *See also* Navigation devices; *individual navigators*
NCs (Curtiss flying boats), 23; *NC-3, 23*
Nelson, Lieutenant Erik H., 42, *43;* quoted, 45
New Orleans (Douglas World Cruiser), *40,* 41, *45*
New South Wales, Australia, 122, 123
New York, *map* 96-97, *map* 170-171; *Daily Mirror,* 85; first nonstop flight from Paris, 139; Lindbergh and, 81, 83, 84-86, 88-89, 97, *102-105;* Orteig Prize and, 51-52, 53, 54, 64, 79, 80, 81; and round-the-world flights, 139, 146, 151, 155, 157, 159; *Times,* 86, 105, 159; *World,* 140
Newfoundland, 17, 19-22, 24-26, 30-33, 39; Lindbergh and, 92, 95; and Orteig Prize, 80, 81; Post and, 146, *147. See also* St. John's, Newfoundland
Nieuport (French plane), 77
Northcliffe, Lord Alfred, *16;* contest rules, *17;* prize offered by, 15-*39,* 51
Nungesser, Lieutenant Charles Eugène Jules Marie, 64, 67, 71, 77-*78, 81;* disappearance of, 72, 81, *82,* 85-86, 96; quoted, 78, 79, 80. *See also L'Oiseau Blanc*
Nungesser, Madame Laure, *82*

O
Oakland, Calif., 118, 125, 128, *map* 170-171
Ogden, Sergeant Henry H., *42,* 47
Orion-Explorer (Lockheed hybrid), 160; crash of, *161*
Orteig, Raymond, 51, 53, 97, 105
Orteig Prize, 51-54, 63-64, 67-71, *105,* 139; Lindbergh and, 63-64, 67, 69-71, 81, 83, 95, 97; Nungesser and, 64, 67, 71, 77, 79-81; planes contending for, *72-75*

P
Pacific Ocean, *map* 170-171; first nonstop flight across, 144-*145;* flights across, 118, 119, 121, 122-123, 125-127, 129, 139, 161
Page, Ray, 56
Pan American Airways, 161
Pangborn, Clyde, 144-*145*
Paris, France, *map* 96-97, *map* 170-171; American Legion Prize and, 65; Lindbergh and, *front endpaper,* 92, 95-9, 97; to New York, first nonstop flight, 139; Nungesser and, 78, 79, 80-81; Orteig Prize and, 51, 52, 54, 64; and U.S. World Cruise, 46
Patrick, General Mason M., 40
Paulhan, Louis, 16

Periscope, in Lindbergh's plane, 66
Peruvian Air Force, motto of, 9
Petit Journal, Le, cover of, 6-7
Pilot training: British, 16, 18; U.S., 55-56, 60-61
Pisa, Italy, 112, *map* 170-171
PL-8 (Levasseur plane). *See L'Oiseau Blanc*
PN-9 flying boats (U.S. Navy), 118
Polando, John, 139
Post, Gordon, quoted, 140-141
Post, Mae Laine, 142, *154;* quoted, 157
Post, Wiley, *138,* 140-144, 146-147, 150-151, *154, 158,* 162, *map* 170-171; death of, 160-*161;* and high-altitude flying, 159-160; license of, *142-143;* quoted, 140, 141, 142, 144, 146, 150, 151, 154; solo flight, 154-157, 159, 161. *See also Winnie Mae*
Poulet, Étienne, 109, 110, 111, 112, 114
Pratt & Whitney company. *See* Wasp engine
Pressure suits, 159-160
Prizes: American Legion, 64; *Asahi Shimbun* (Japanese newspaper), 144; Australia, 110-114; Collier Trophy, 55; Northcliffe, 15-*39,* 51. *See also* Orteig Prize
Punch (British magazine), cartoon from, *118*

Q

Quidi Vidi lake (Newfoundland), 20, 24, 25, 26, 30; Vimy assembled at, *27-29*

R

Rangoon, Burma, 113, 114, *map* 170-171
Raymor (Martinsyde plane), 19, 22, *24,* 26, 30
Raynham, Frederick "Freddie," 18-22, *21,* 24-25, 30, 33
Read, Lieutenant Commander Albert C., *map* 96-97; quoted, 23
Rendle, Valentine, 119, *122*
Reynolds, William J., painting by, *90;* quoted, 90
Richmond, U.S.S. (U.S. ship), launch from, *48*
Robertson, William and Frank, 61, 63, 65
Robertson Aircraft Corporation, 61-62, 63, 65; D.H.4, *62*
Roe, A. V., and Company, 116
Rogers, Will, *160;* death of, 160-*161;* quoted, 160
Rolls-Royce engines, 27, 30; Eagle, 19; Eagle Mark VIII, 17, 18; Falcon, 19
Rome, Italy, 116, *map* 170-171; Ostia port, *136-137*
Roosevelt Field (Long Island, N.Y.), *52-53,* 146, *154;* Lindbergh and, *76,* 86, 88, *89,* 97
Round-the-world flights, *40-41,* 139-161; first solo, 154-157, 159; *Graf Zeppelin,* 140, *141,* 151; Hughes's, *162-169;* U.S. World Cruise, *40-49, map* 41
Royal Aero Club (London), 110, 114
Royal Air Force (RAF), 17, 97, 110, 115, 121, 122, 139; Sealand Aerodrome, 146
Royal Naval Air Service, 17, 19, 109
Russia, and round-the-world flights, 146-147, 150, 155, 156-157, *map* 170-171
Ryan, T. Claude, 65-66; quoted, 83
Ryan Airlines, 65-66, 67, 69, 70, 79, 81, 83, 143; Ryan M-2 (monoplane), 65-67. *See also Spirit of St. Louis*

S

St. John's, Newfoundland, 20-22, *21,* 24-26, 31, 33, *map* 96-97; Cochrane Hotel, 20-*21,* 22, 32, 36; Lester's Field, 30, *31, 32;* Lindbergh and, 92, 96
St. Louis, Mo., Lindbergh and, 61, 63, 65, 81, 83, 84, *map* 170-171
St. Raphael (Fokker plane), 64
San Diego, Calif., 51, 81, 118, *map* 170-171. *See also* Ryan Airlines
San Francisco, Calif., 51-52, *map* 170-171
Santa Monica, Calif., Clover Field, *40-41*
Savoia-Marchetti Sm. 55X flying boats, *130-137*
Seaplanes: Douglas World Cruisers, *40-49;* Orion-Explorer, 160-*161;* Short Brothers, 19. *See also* Flying boats
Seattle (Douglas World Cruiser), 40, 43, *44*
Shamrock (Wood's plane), *18-19*
Shiers, Sergeant Wally, 111, 112, 113; quoted, 114
Short Brothers seaplane, 19
Sikorsky, Igor, 52-*53,* 54; quoted, 6
Sikorsky S-35 (biplane), *52-53,* 54, 63, *72*
Silver Queen (Vickers Vimy), 115
Singapore, 113-114, *map* 170-171
Smith, Reverend and Mrs. Jasper, *49*
Smith, Keith, 111, 114, 116, 117, *map* 170-171
Smith, Lieutenant Lowell H., *42,* 44, 45, *46, 48, 49;* quoted, 48
Smith, Captain Ross, 109, 110, 111-114, *map* 111, 117, *map* 170-171; quoted, 112
Smithsonian Institution, 160
Sonoma, S.S. (U.S. ship), 119
Sopwith Airplane Company, 16, 19; Dove biplane, 109, 114, 116; flying school, 18; Wallaby, 110-111. *See also Atlantic* (Sopwith plane)
South Africa, 139; flight to, *115*
Southern Cross (Fokker Trimotor), 125-127, *128-129*
Sperry, Elmer, Jr., instrument designed by, *149*
Sperry Gyroscope Company, 155
Spirit of St. Louis, 67, 69-*71,* 72, *74-75, 85, 98-99,* 122; cockpit of, *92;* New York to Paris flight, *76, 89-*96, *90-91, 94,* 97; paintings of, *front endpaper, 90-91;* San Diego to New York flight, 81, 83-86, 88-89. *See also* Ryan Airlines
Sunrise (U.S. biplane), 19
Sunstedt, Hugo, 19
Suva, Fiji Islands, 125, 126, 127, 129, 139, *map* 170-171
Sydney, Australia, 126, 127, *128-129, map* 170-171; airport painting, *108*

T

Texas Topnotch Flyers, 141
Tibbs, Burrell, 141; quoted, 141
Timor, 114, *map* 170-171
Trimotors (Fokker planes), 64, 67, 68-69, 118, 123, 125. *See also America* (Fokker Trimotor); *Bird of Paradise* (Fokker Trimotor); Ford Trimotor; *Southern Cross* (Fokker Trimotor)

U

Ulm, Charles T. P., *108,* 119, 121-123, 125-127, 129, *map* 170-171; quoted, 126, 127
United States: civilian aviation in, 52; competition with France, 51-52; first flight from Australia to, 129; flights across, 140, 160; flights to Australia from, 118, 119, 121, 122-123, 125-127, 129; flights to Hawaii from, 118-119; World Cruise, *40-49*
United States Army: Air Corps, 118, 148; Air Service, 40, 60-61, 62; planes of, 51-52; round-the-world flights by, *40-41, map* 41, 139, *map* 170-171
United States Congress, and Lindbergh medal, *100*
United States Navy, 40, 53, 81; Academy of, 64; Curtiss flying boats (NCs), *23;* flights to Hawaii, 118
United States Postal Service, 61, 119

V

Van Ryneveld, Lieutenant Colonel H. A., 115, *map* 170-171; flag to, *115;* medicine chest of, *115*
Védrines, Jules, *12,* 109-110
Verne, Jules, 139-140
Vickers Ltd.: aviation division, 16, 17, 25, 27, 30; Viking, 39. *See also* Vimy (Vickers bomber)
Vimy (Vickers bomber): Alcock and Brown and, 17-20, 25-26, 30-33, *31, 32,* 36-39, *38,* 52; assembling of, *27-29;* radio of, *34;* Smith brothers and, 110, 112-114, 116

W

Wade, Lieutenant Leigh, *42,* 47; quoted, 40
Walker, James J., quoted, 105
Wanamaker, Rodman, 64
Warner, James, 125, 129
Washington, D.C., *map* 170-171; Bolling Field, *49;* Lindbergh at, *100-101;* Smithsonian Institution, 160
Wasp engine (Pratt & Whitney), 143, 144, 146, *153,* 160
Wells, Linton, *45,* 140
Wenatchee, Wash., *map* 170-171; museum exhibits in, 144-*145*
Whirlwind engines (Wright), 65-66, 67, 69-70, 143; advertisement for, *54;* and Australia flights, 118, 123, 125; J-5C, *54-55,* 67; Lindbergh and, 86, 92, 95; and Orteig Prize planes, 72, 73
Wiley Post Institute for Aeronautical Research, 151, 154
Winnie Mae (Lockheed Vega), *138,* 143-144, 146-*147,* 150-151, *152-153;* and high-altitude flying, 159-161; and Post's solo flight, 155-157, 159
Wood, Major J. C. P., 19; quoted, 18; rescue of, *18-19*
Wooster, Lieutenant Stanton Hall, 67; death of, *68*
World War I, 6, 17; Allied aces, 52, *53,* 64, 77-*78, 81;* British planes in, 16; peacetime aviation following, 15; Versailles treaty and, 110
World War II, 161
Wright, Orville, 54; and Post's license, *142-143*
Wright Aeronautical Corporation, 54, 64, 65; and Lindbergh, 84, 86. *See also* Cyclone engines (Wright); Whirlwind engines (Wright); Wright-Bellanca monoplanes
Wright-Bellanca monoplanes, 54-55, 63-64, 65, 139; *Miss Veedol, 144-145. See also Columbia* (Wright-Bellanca monoplane)

Y

Yangtze River, *45*

Printed in U.S.A.